MULTIHAZARD RISK ATLAS OF MALDIVES

Climate and Geophysical Hazards—Volume II

MARCH 2020

 Creative Commons Attribution 3.0 IGO license (CC BY 3.0 IGO)

© 2020 Asian Development Bank
6 ADB Avenue, Mandaluyong City, 1550 Metro Manila, Philippines
Tel +63 2 8632 4444; Fax +63 2 8636 2444
www.adb.org

Some rights reserved. Published in 2020.

ISBN 978-92-9262-045-5 (print); 978-92-9262-046-2 (electronic); 978-92-9262-047-9 (ebook)
Publication Stock No. TCS200050
DOI: http://dx.doi.org/10.22617/TCS200050

The views expressed in this publication are those of the authors and do not necessarily reflect the views and policies of the Asian Development Bank (ADB) or its Board of Governors or the governments they represent.

ADB does not guarantee the accuracy of the data included in this publication and accepts no responsibility for any consequence of their use. The mention of specific companies or products of manufacturers does not imply that they are endorsed or recommended by ADB in preference to others of a similar nature that are not mentioned.

By making any designation of or reference to a particular territory or geographic area, or by using the term "country" in this document, ADB does not intend to make any judgments as to the legal or other status of any territory or area.

This work is available under the Creative Commons Attribution 3.0 IGO license (CC BY 3.0 IGO) https://creativecommons.org/licenses/by/3.0/igo/. By using the content of this publication, you agree to be bound by the terms of this license. For attribution, translations, adaptations, and permissions, please read the provisions and terms of use at https://www.adb.org/terms-use#openaccess.

This CC license does not apply to non-ADB copyright materials in this publication. If the material is attributed to another source, please contact the copyright owner or publisher of that source for permission to reproduce it. ADB cannot be held liable for any claims that arise as a result of your use of the material.

Please contact pubsmarketing@adb.org if you have questions or comments with respect to content, or if you wish to obtain copyright permission for your intended use that does not fall within these terms, or for permission to use the ADB logo.

Corrigenda to ADB publications may be found at http://www.adb.org/publications/corrigenda.

Notes:
In this publication, "$" refers to United States dollars.
The maps presented in this atlas reflect airports based on 2017 data from the Civil Aviation Authority of Maldives.

On the cover: An aerial view shows 1 of 26 natural atolls that make up Maldives, which also includes nearly 1,200 small coral islands and some of the world's most beautiful beaches. Recognized as the seventh-largest in the world, the coral reefs and associated ecosystems of Maldives are key foundations for food security and means of livelihood. Yet, they are considered as among the most vulnerable to climate change (photo by Roberta Gerpacio).

Contents
iii

Tables and Maps	v
Foreword	ix
Acknowledgments	x
Abbreviations	xi
Climate	**1**
Dynamic Climate	**2**
Automatic Weather Stations and Meteorological Observation Stations	**3**
Historical Annual Climate	**17**
Rainfall	17
Temperature	17
Historical Seasonal Climate	**20**
Seasonal Average Rainfall (1970–2005)	21
Seasonal Average Temperature (1970–2005)	26
Future Climate	**31**
Average Annual Rainfall Projection (RCP 4.5)	32
Average Seasonal Rainfall Projection (DJF, RCP 4.5)	37
Average Seasonal Rainfall Projection (MAM, RCP 4.5)	42
Average Seasonal Rainfall Projection (JJA, RCP 4.5)	47
Average Seasonal Rainfall Projection (SON, RCP 4.5)	52
Average Annual Rainfall Projection (RCP 8.5)	57
Average Seasonal Rainfall Projection (DJF, RCP 8.5)	62
Average Seasonal Rainfall Projection (MAM, RCP 8.5)	67

Average Seasonal Rainfall Projection (JJA, RCP 8.5)	72
Average Seasonal Rainfall Projection (SON, RCP 8.5)	77
Average Annual Temperature Projection (RCP 4.5)	82
Average Seasonal Temperature Projection (DJF, RCP 4.5)	87
Average Seasonal Temperature Projection (MAM, RCP 4.5)	92
Average Seasonal Temperature Projection (JJA, RCP 4.5)	97
Average Seasonal Temperature Projection (SON, RCP 4.5)	102
Average Annual Temperature Projection (RCP 8.5)	107
Average Seasonal Temperature Projection (DJF, RCP 8.5)	112
Average Seasonal Temperature Projection (MAM, RCP 8.5)	117
Average Seasonal Temperature Projection (JJA, RCP 8.5)	122
Average Seasonal Temperature Projection (SON, RCP 8.5)	127
Summary of Observations for Rainfall	**132**
Summary of Observations for Temperature	**133**
Geophysical Hazards	**134**
Map Data Sources	**139**
References	**140**

Tables and Maps

Tables

II.1	Islands with Weather Stations	3
II.2	Variation in Rainfall Patterns across Space, Time, and Representative Concentration Pathways	132
II.3	Variation in Temperature Patterns across Space, Time, and Representative Concentration Pathways	133

Maps

II.1	Maldives, Meteorological Observation Stations	4
II.2	Addu City, Meteorological Observation Stations	5
II.3	Alifu Alifu Atoll, Meteorological Observation Stations	6
II.4	Alifu Dhaalu Atoll, Meteorological Observation Stations	7
II.5	Gaafu Alifu Atoll, Meteorological Observation Stations	8
II.6	Gaafu Dhaalu Atoll, Meteorological Observation Stations	9
II.7	Haa Alifu Atoll, Meteorological Observation Stations	10
II.8	Haa Dhaalu Atoll, Meteorological Observation Stations	11
II.9	Laamu Atoll, Meteorological Observation Stations	12
II.10	North Malé Atoll, Meteorological Observation Stations	13
II.11	Raa Atoll, Meteorological Observation Stations	14
II.12	Thaa Atoll, Meteorological Observation Stations	15
II.13	Vaavu Atoll, Meteorological Observation Stations	16
II.14	Maldives, Annual Average Rainfall (1970–2005)	18
II.15	Maldives, Annual Average Temperature (1970–2005)	19
II.16	Maldives, Seasonal Average Rainfall (1970–2005)	21
II.17	Maldives, Seasonal Average Rainfall (DJF 1970–2005)	22

II.18	Maldives, Seasonal Average Rainfall (MAM 1970–2005)	23
II.19	Maldives, Seasonal Average Rainfall (JJA 1970–2005)	24
II.20	Maldives, Seasonal Average Rainfall (SON 1970–2005)	25
II.21	Maldives, Seasonal Average Temperature (1970–2005)	26
II.22	Maldives, Seasonal Average Temperature (DJF 1970–2005)	27
II.23	Maldives, Seasonal Average Temperature (MAM 1970–2005)	28
II.24	Maldives, Seasonal Average Temperature (JJA 1970–2005)	29
II.25	Maldives, Seasonal Average Temperature (SON 1970–2005)	30
II.26	Maldives, Annual Average Rainfall Projection (RCP 4.5)	32
II.27	Maldives, Annual Average Rainfall (2011–2020, RCP 4.5)	33
II.28	Maldives, Annual Average Rainfall (2021–2030, RCP 4.5)	34
II.29	Maldives, Annual Average Rainfall (2031–2040, RCP 4.5)	35
II.30	Maldives, Annual Average Rainfall (2041–2050, RCP 4.5)	36
II.31	Maldives, Average Seasonal Rainfall Projection (DJF, RCP 4.5)	37
II.32	Maldives, Seasonal Average Rainfall (DJF 2011–2020, RCP 4.5)	38
II.33	Maldives, Seasonal Average Rainfall (DJF 2021–2030, RCP 4.5)	39
II.34	Maldives, Seasonal Average Rainfall (DJF 2031–2040, RCP 4.5)	40
II.35	Maldives, Seasonal Average Rainfall (DJF 2041–2050, RCP 4.5)	41
II.36	Maldives, Average Seasonal Rainfall Projection (MAM, RCP 4.5)	42
II.37	Maldives, Seasonal Average Rainfall (MAM 2011–2020, RCP 4.5)	43
II.38	Maldives, Seasonal Average Rainfall (MAM 2021–2030, RCP 4.5)	44
II.39	Maldives, Seasonal Average Rainfall (MAM 2031–2040, RCP 4.5)	45
II.40	Maldives, Seasonal Average Rainfall (MAM 2041–2050, RCP 4.5)	46
II.41	Maldives, Average Seasonal Rainfall Projection (JJA, RCP 4.5)	47
II.42	Maldives, Seasonal Average Rainfall (JJA 2011–2020, RCP 4.5)	48
II.43	Maldives, Seasonal Average Rainfall (JJA 2021–2030, RCP 4.5)	49
II.44	Maldives, Seasonal Average Rainfall (JJA 2031–2040, RCP 4.5)	50
II.45	Maldives, Seasonal Average Rainfall (JJA 2041–2050, RCP 4.5)	51
II.46	Maldives, Average Seasonal Rainfall Projection (SON, RCP 4.5)	52
II.47	Maldives, Seasonal Average Rainfall (SON 2011–2020, RCP 4.5)	53
II.48	Maldives, Seasonal Average Rainfall (SON 2021–2030, RCP 4.5)	54
II.49	Maldives, Seasonal Average Rainfall (SON 2031–2040, RCP 4.5)	55
II.50	Maldives, Seasonal Average Rainfall (SON 2041–2050, RCP 4.5)	56
II.51	Maldives, Average Annual Rainfall Projection (RCP 8.5)	57
II.52	Maldives, Annual Average Rainfall (2011–2020, RCP 8.5)	58
II.53	Maldives, Annual Average Rainfall (2021–2030, RCP 8.5)	59
II.54	Maldives, Annual Average Rainfall (2031–2040, RCP 8.5)	60
II.55	Maldives, Annual Average Rainfall (2041–2050, RCP 8.5)	61
II.56	Maldives, Average Seasonal Rainfall Projection (DJF, RCP 8.5)	62

II.57	Maldives, Seasonal Average Rainfall (DJF 2011–2020, RCP 8.5)	63
II.58	Maldives, Seasonal Average Rainfall (DJF 2021–2030, RCP 8.5)	64
II.59	Maldives, Seasonal Average Rainfall (DJF 2031–2040, RCP 8.5)	65
II.60	Maldives, Seasonal Average Rainfall (DJF 2041–2050, RCP 8.5)	66
II.61	Maldives, Average Seasonal Rainfall Projection (MAM, RCP 8.5)	67
II.62	Maldives, Seasonal Average Rainfall (MAM 2011–2020, RCP 8.5)	68
II.63	Maldives, Seasonal Average Rainfall (MAM 2021–2030, RCP 8.5)	69
II.64	Maldives, Seasonal Average Rainfall (MAM 2031–2040, RCP 8.5)	70
II.65	Maldives, Seasonal Average Rainfall (MAM 2041–2050, RCP 8.5)	71
II.66	Maldives, Average Seasonal Rainfall Projection (JJA, RCP 8.5)	72
II.67	Maldives, Seasonal Average Rainfall (JJA 2011–2020, RCP 8.5)	73
II.68	Maldives, Seasonal Average Rainfall (JJA 2021–2030, RCP 8.5)	74
II.69	Maldives, Seasonal Average Rainfall (JJA 2031–2040, RCP 8.5)	75
II.70	Maldives, Seasonal Average Rainfall (JJA 2041–2050, RCP 8.5)	76
II.71	Maldives, Average Seasonal Rainfall Projection (SON, RCP 8.5)	77
II.72	Maldives, Seasonal Average Rainfall (SON 2011–2020, RCP 8.5)	78
II.73	Maldives, Seasonal Average Rainfall (SON 2021–2030, RCP 8.5)	79
II.74	Maldives, Seasonal Average Rainfall (SON 2031–2040, RCP 8.5)	80
II.75	Maldives, Seasonal Average Rainfall (SON 2041–2050, RCP 8.5)	81
II.76	Maldives, Average Annual Temperature Projection (RCP 4.5)	82
II.77	Maldives, Annual Average Temperature (2011–2020, RCP 4.5)	83
II.78	Maldives, Annual Average Temperature (2021–2030, RCP 4.5)	84
II.79	Maldives, Annual Average Temperature (2031–2040, RCP 4.5)	85
II.80	Maldives, Annual Average Temperature (2041–2050, RCP 4.5)	86
II.81	Maldives, Average Seasonal Temperature Projection (DJF, RCP 4.5)	87
II.82	Maldives, Seasonal Average Temperature (DJF 2011–2020, RCP 4.5)	88
II.83	Maldives, Seasonal Average Temperature (DJF 2021–2030, RCP 4.5)	89
II.84	Maldives, Seasonal Average Temperature (DJF 2031–2040, RCP 4.5)	90
II.85	Maldives, Seasonal Average Temperature (DJF 2041–2050, RCP 4.5)	91
II.86	Maldives, Average Seasonal Temperature Projection (MAM, RCP 4.5)	92
II.87	Maldives, Seasonal Average Temperature (MAM 2011–2020, RCP 4.5)	93
II.88	Maldives, Seasonal Average Temperature (MAM 2021–2030, RCP 4.5)	94
II.89	Maldives, Seasonal Average Temperature (MAM 2031–2040, RCP 4.5)	95
II.90	Maldives, Seasonal Average Temperature (MAM 2041–2050, RCP 4.5)	96
II.91	Maldives, Average Seasonal Temperature Projection (JJA, RCP 4.5)	97
II.92	Maldives, Seasonal Average Temperature (JJA 2011–2020, RCP 4.5)	98
II.93	Maldives, Seasonal Average Temperature (JJA 2021–2030, RCP 4.5)	99
II.94	Maldives, Seasonal Average Temperature (JJA 2031–2040, RCP 4.5)	100
II.95	Maldives, Seasonal Average Temperature (JJA 2041–2050, RCP 4.5)	101

II.96	Maldives, Average Seasonal Temperature Projection (SON, RCP 4.5)	102
II.97	Maldives, Seasonal Average Temperature (SON 2011–2020, RCP 4.5)	103
II.98	Maldives, Seasonal Average Temperature (SON 2021–2030, RCP 4.5)	104
II.99	Maldives, Seasonal Average Temperature (SON 2031–2040, RCP 4.5)	105
II.100	Maldives, Seasonal Average Temperature (SON 2041–2050, RCP 4.5)	106
II.101	Maldives, Average Annual Temperature Projection (RCP 8.5)	107
II.102	Maldives, Annual Average Temperature (2011–2020, RCP 8.5)	108
II.103	Maldives, Annual Average Temperature (2021–2030, RCP 8.5)	109
II.104	Maldives, Annual Average Temperature (2031–2040, RCP 8.5)	110
II.105	Maldives, Annual Average Temperature (2041–2050, RCP 8.5)	111
II.106	Maldives, Average Seasonal Temperature Projection (DJF, RCP 8.5)	112
II.107	Maldives, Seasonal Average Temperature (DJF 2011–2020, RCP 8.5)	113
II.108	Maldives, Seasonal Average Temperature (DJF 2021–2030, RCP 8.5)	114
II.109	Maldives, Seasonal Average Temperature (DJF 2031–2040, RCP 8.5)	115
II.110	Maldives, Seasonal Average Temperature (DJF 2041–2050, RCP 8.5)	116
II.111	Maldives, Average Seasonal Temperature Projection (MAM, RCP 8.5)	117
II.112	Maldives, Seasonal Average Temperature (MAM 2011–2020, RCP 8.5)	118
II.113	Maldives, Seasonal Average Temperature (MAM 2021–2030, RCP 8.5)	119
II.114	Maldives, Seasonal Average Temperature (MAM 2031–2040, RCP 8.5)	120
II.115	Maldives, Seasonal Average Temperature (MAM 2041–2050, RCP 8.5)	121
II.116	Maldives, Average Seasonal Temperature Projection (JJA, RCP 8.5)	122
II.117	Maldives, Seasonal Average Temperature (JJA 2011–2020, RCP 8.5)	123
II.118	Maldives, Seasonal Average Temperature (JJA 2021–2030, RCP 8.5)	124
II.119	Maldives, Seasonal Average Temperature (JJA 2031–2030, RCP 8.5)	125
II.120	Maldives, Seasonal Average Temperature (JJA 2041–2050, RCP 8.5)	126
II.121	Maldives, Average Seasonal Temperature Projection (SON, RCP 8.5)	127
II.122	Maldives, Seasonal Average Temperature (SON 2011–2020, RCP 8.5)	128
II.123	Maldives, Seasonal Average Temperature (SON 2021–2030, RCP 8.5)	129
II.124	Maldives, Seasonal Average Temperature (SON 2031–2040, RCP 8.5)	130
II.125	Maldives, Seasonal Average Temperature (SON 2041–2050, RCP 8.5)	131
II.126	Maldives, Cyclonic Wind Hazard Zone	135
II.127	Maldives, Surge Hazard Zone	136
II.128	Maldives, Seismic Hazard Zone	137
II.129	Maldives, Tsunami Hazard Zone	138

Foreword

Maldives is among the countries most vulnerable to the impacts of climate change as it is a small island nation with extremely low elevations. Maldives is also very vulnerable to impacts of rising air and sea surface temperatures and changes in rainfall patterns. Climate change impacts will therefore impose significant negative consequences on the Maldivian economy and society. Some of the priority vulnerabilities to climate change are land loss and beach erosion, infrastructure damage, degradation of coral reefs, and adverse impacts on water resources, food security, human health, and the overall economy.

Sustainable coastal resources management is of particular importance to Maldives, such that all regulations involving various development activities have coastal components. Despite the government's continued efforts in improving and sustaining coastal resources management, critical issues remain, such as the need for systematized coastal monitoring, clear definition of coastal boundaries and coastal development, enhanced regulatory and monitoring capacities for coastal resources protection, and sustainable long-term strategies on land reclamation and marine area protection. At a time when climate is rapidly changing and extreme weather events are frequently occurring, the critical roles that marine and coastal environments play in mitigating and adapting to climate change need to be sufficiently documented and properly recognized. It is therefore essential for Maldives to develop and establish a comprehensive digital database of marine and coastal ecosystem features and services that can be regularly monitored.

The *Multihazard Risk Atlas of Maldives* was developed through the project "Establishing a National Geospatial Database for Mainstreaming Climate Change Adaptation into Development Activities and Policies in Maldives" under the Asian Development Bank's regional knowledge and support (capacity development) technical assistance Action on Climate Change in South Asia (2013–2018). This five-volume atlas aims to promote the sustainable development of coastal and marine ecosystems and their various components, by enhancing the awareness of stakeholders on and enjoining them to address climate and disaster risks (including hazards, exposures, and vulnerabilities) to which ecosystems are exposed. The atlas presents spatial information and maps necessary for assessing future development investments in terms of their risks to climate and geophysical hazards.

The target audience of the *Multihazard Risk Atlas of Maldives* are the concerned stakeholders with current or planned development activities in the country, including public and private sectors, nongovernment organizations, research and academic community, development partner agencies, other financial institutions, and the general public. The atlas will also be a useful reference for other developing countries with similar geographical and environmental conditions, particularly small island developing states. It is envisioned that the atlas will significantly contribute to rendering important sector development investments more resilient to hazard-specific risk scenarios in the short, medium, and long terms.

H.E. Dr. Hussain Rasheed Hassan
Minister
Ministry of Environment, Malé

Shixin Chen
Vice-President for Operations 1
Asian Development Bank, Manila

Acknowledgments

Government Ministries, Departments, and Agencies in Maldives
    Civil Aviation Authority
    Land and Survey Authority
    Marine Research Institute
    Meteorological Service
    Ministry of Economic Development
    Ministry of Education
    Ministry of Environment
    Ministry of Fisheries, Marine Resources and Agriculture
    Ministry of Health
    Ministry of National Planning and Infrastructure
    Ministry of Tourism
    National Bureau of Statistics
    National Disaster Management Center

International Institutions
    Manila Observatory
    Marine Spatial Ecology Lab, University of Queensland, Australia
    SANDER + PARTNER
    United Nations Development Programme

International Institutions in Maldives
    International Union for Conservation of Nature, Maldives
    United Nations Development Programme, Maldives

National Consultant Team
    Ahmed Jameel, Integrated Coastal Zone Management Specialist
    Faruhath Jameel, Geographic Information Systems Specialist and Team Leader
    Hussain Naeem, Coastal Ecosystems and Biodiversity Specialist
    Mahmood Riyaz, Climate Change Risk Assessment Specialist

Abbreviations

°C	–	Celsius
AWS	–	Automated Weather Station
BODC	–	British Oceanographic Data Centre
CAA	–	Maldives Civil Aviation Authority
DJF	–	December, January, February
GEBCO	–	General Bathymetric Chart of the Oceans
GHCN	–	Global Historical Climatology Network
IHO	–	International Hydrographic Organization
IOC	–	Intergovernmental Oceanographic Commission
JJA	–	June, July, August
MAM	–	March, April, May
ME	–	Ministry of Environment
MED	–	Ministry of Economic Development
MLSA	–	Maldives Land and Survey Authority
mm/day	–	millimeter per day
MMS	–	Maldives Meteorological Service
RCP	–	Representative Concentration Pathway
SON	–	September, October, November
UNDP	–	United Nations Development Programme
UTM	–	Universal Transverse Mercator
WGS	–	World Geodetic System

Climate 1

This volume contains maps of the locations of weather monitoring stations and the historical climate and projected climate (average rainfall and temperature) based on Representative Concentration Pathways (RCP) 4.5 and 8.5. Only the projected actual average annual and seasonal rainfall and temperature values were included in this risk atlas as these are simpler to interpret.

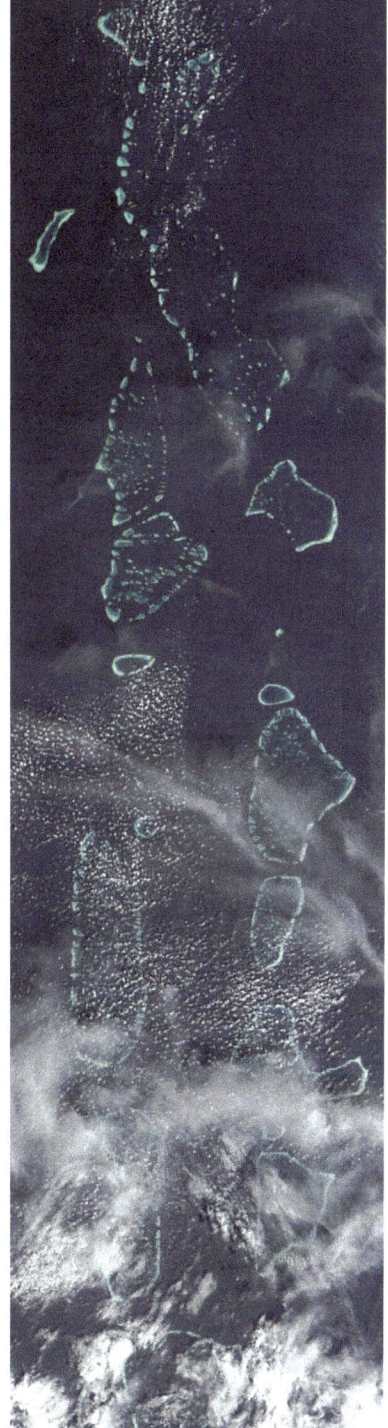

Clouds over Maldives. A satellite image showing dispersed clouds (photo by Jeff Schmaltz, Moderate Resolution Imaging Spectroradiometer Rapid Response Team, NASA's Goddard Space Flight Center).

Dynamic Climate

Maldivians enjoy mostly sunny days. Warm temperature (ranging from 26°C to 33°C) is experienced throughout the year, with little variation, for the entire stretch of Maldives (Moosa 2014).

While temperature stays relatively constant, the rainfall varies depending on the monsoon. Maldives has two monsoon seasons: (i) northeast monsoon or dry season, starting in November and lasting until April; and (ii) southwest monsoon, which brings rainfall from May to October. Rainfall varies across the year and latitude (Moosa 2014).

In Maldives, climate is an important element shaping present and future policies, actions, and decisions. It affects Maldives' coastal and marine resources (Asian Development Bank 2015). Increased global temperature can raise sea levels and inundate the low-lying islands of the country (Khan et al. 2002). Furthermore, the corals would also be damaged by an increase in global temperatures. Water resources as well as agriculture also rely on sufficient precipitation to meet the water and food requirements of the communities. Natural flora and fauna require moderate climate to survive.

To prevent losses due to weather- and climate-related disaster events, it is important to monitor the weather parameters and evaluate the climatic trends in the country. This is done through data gathered from the automatic weather stations and meteorological observation stations scattered in different islands.

Rainbow in horizon. A beautiful rainbow visible in the horizon of a beach in Maldives. Rainbows are formed when there is a refraction and reflection of light in cloud droplets, forming a spectrum of light (photo by Michael Pfütze).

Automatic Weather Stations and Meteorological Observation Stations

Maldives has a total of 18 weather monitoring stations, 5 of which are meteorological observation stations (Table II.1). The locations of the weather monitoring stations are concentrated in the southern region, particularly in Addu and Gaafu Dhaalu atolls. Most of the meteorological observation stations are located in the capital islands and international airports (Map II.1).

Table II.1: Islands with Weather Stations

ATOLL	Island	Station
Addu City	Hithadhoo	Automatic Weather Station
Addu City	Gan	Meteorological Observation Station
Addu City	Meedhoo	Automatic Weather Station
Alifu Alifu	Rasdhoo	Automatic Weather Station
Alifu Dhaalu	Rangali	Automatic Weather Station
Gaafu Alifu	Villingili	Automatic Weather Station
Gaafu Dhaalu	Kaadedhdhoo	Meteorological Observation Station
Gaafu Dhaalu	Gahdhoo	Automatic Weather Station
Gaafu Dhaalu	Rodhavarrehaa	Automatic Weather Station
Haa Alifu	Uligamu	Automatic Weather Station
Haa Dhaalu	Hanimaadhoo	Meteorological Observation Station
Laamu	Isdhoo	Automatic Weather Station
Laamu	Kadhdhoo	Meteorological Observation Station
North Malé	Banyan Tree	Automatic Weather Station
North Malé	Malé	Meteorological Observation Station
Raa	Vadhoo	Automatic Weather Station
Thaa	Hirilandhoo	Automatic Weather Station
Vaavu	Fulidhoo	Automatic Weather Station

Source: Maldives Meteorological Service, 2017.

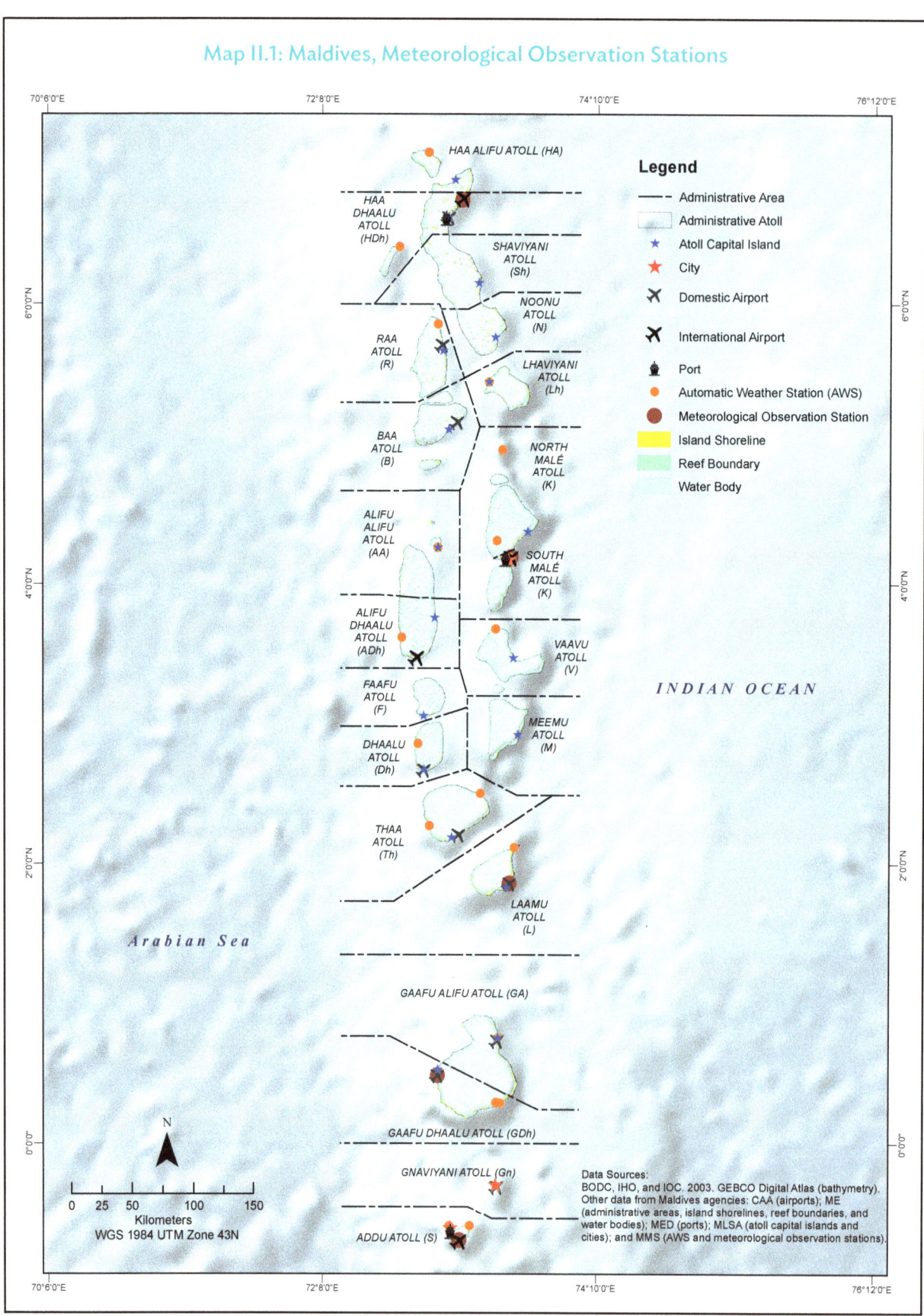

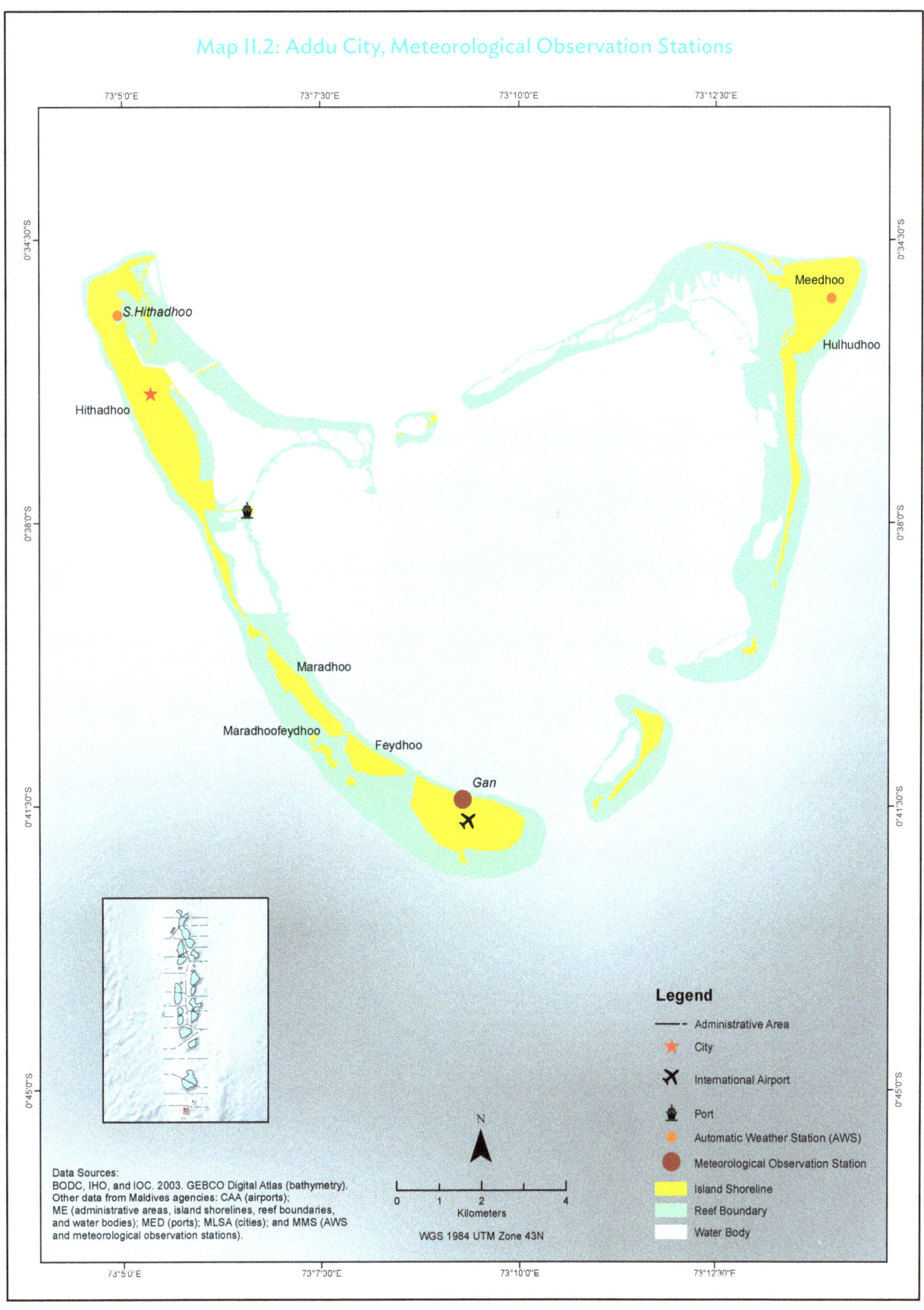

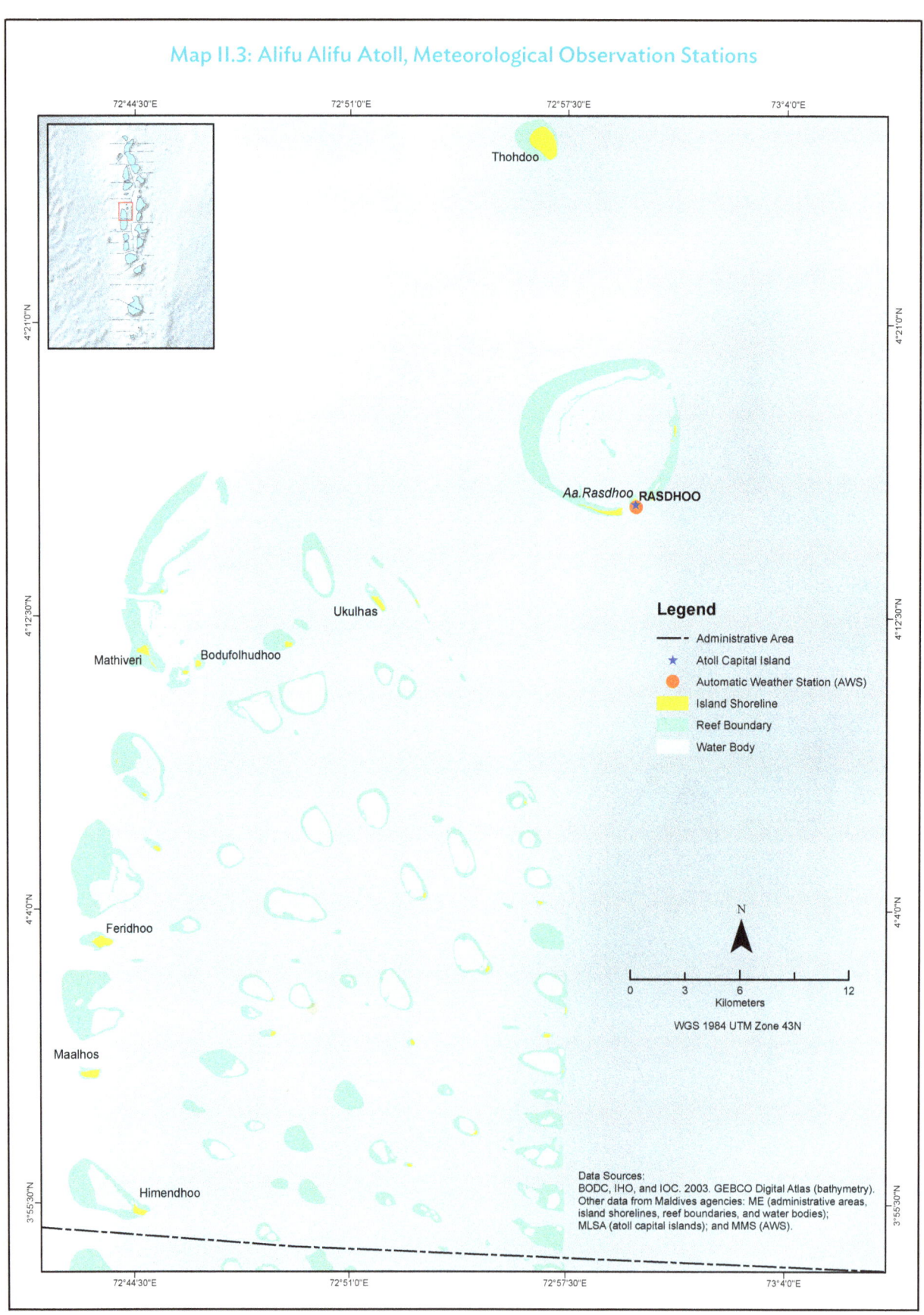

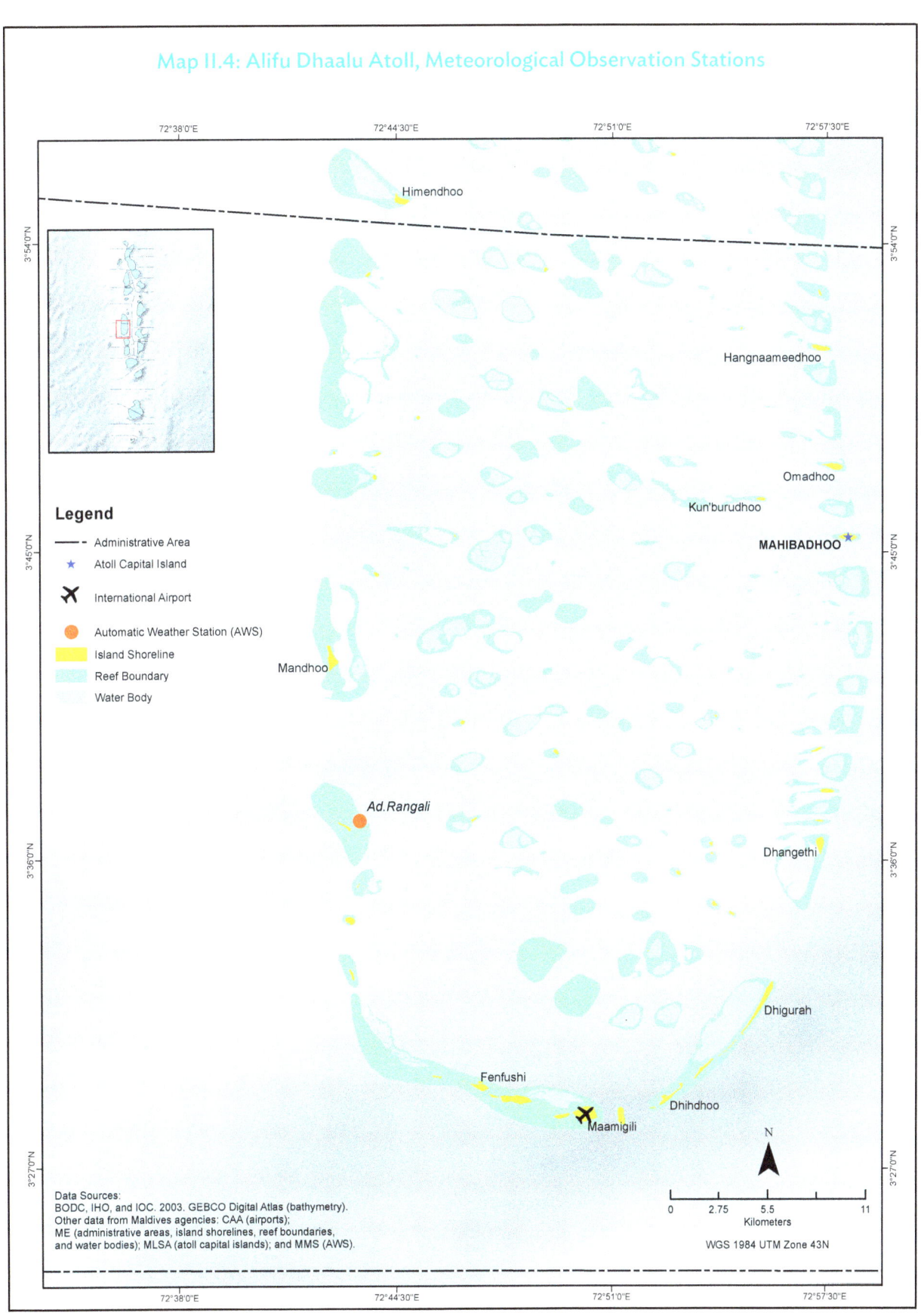

8 Multihazard Risk Atlas of Maldives—Climate and Geophysical Hazards

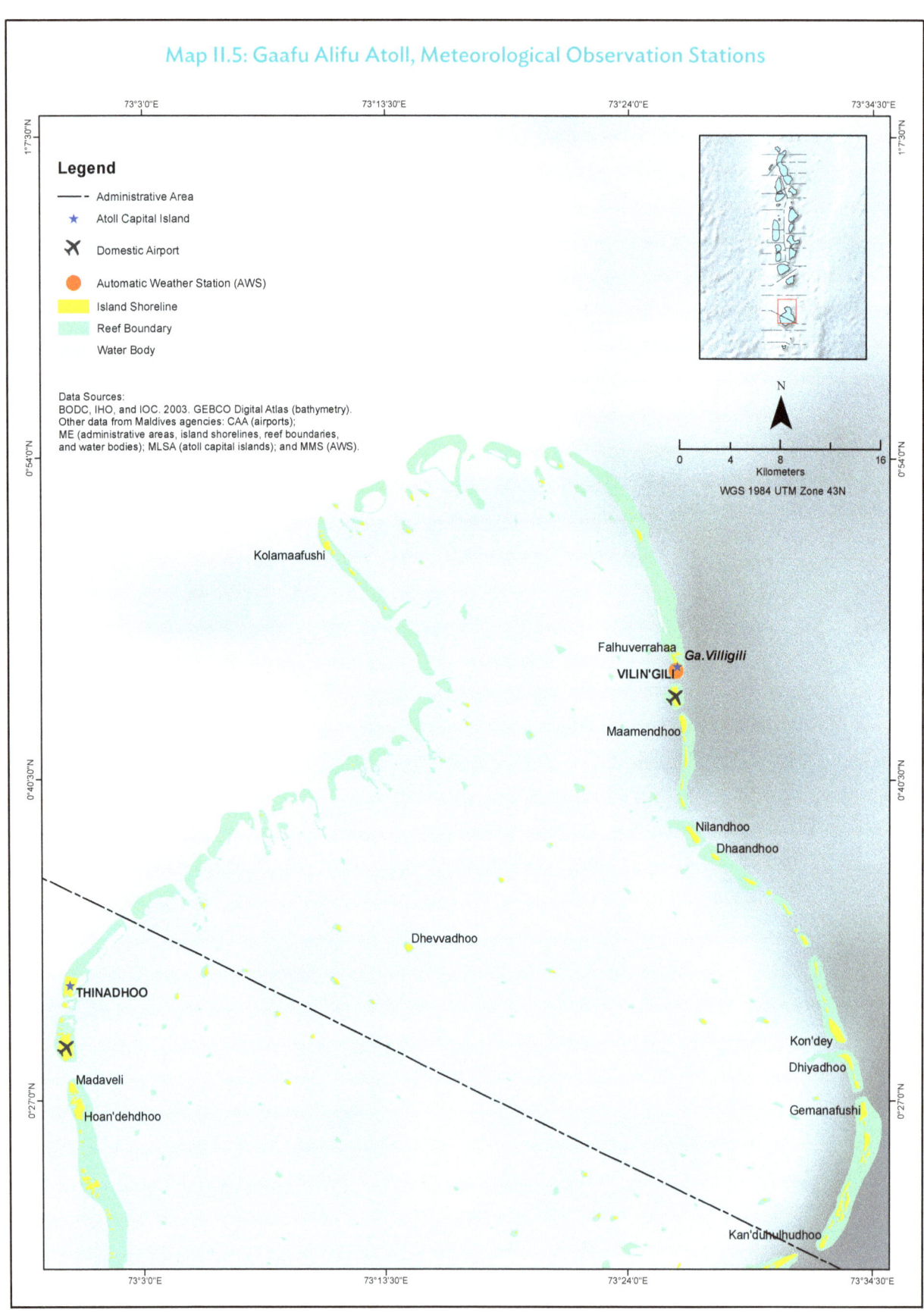

Map II.5: Gaafu Alifu Atoll, Meteorological Observation Stations

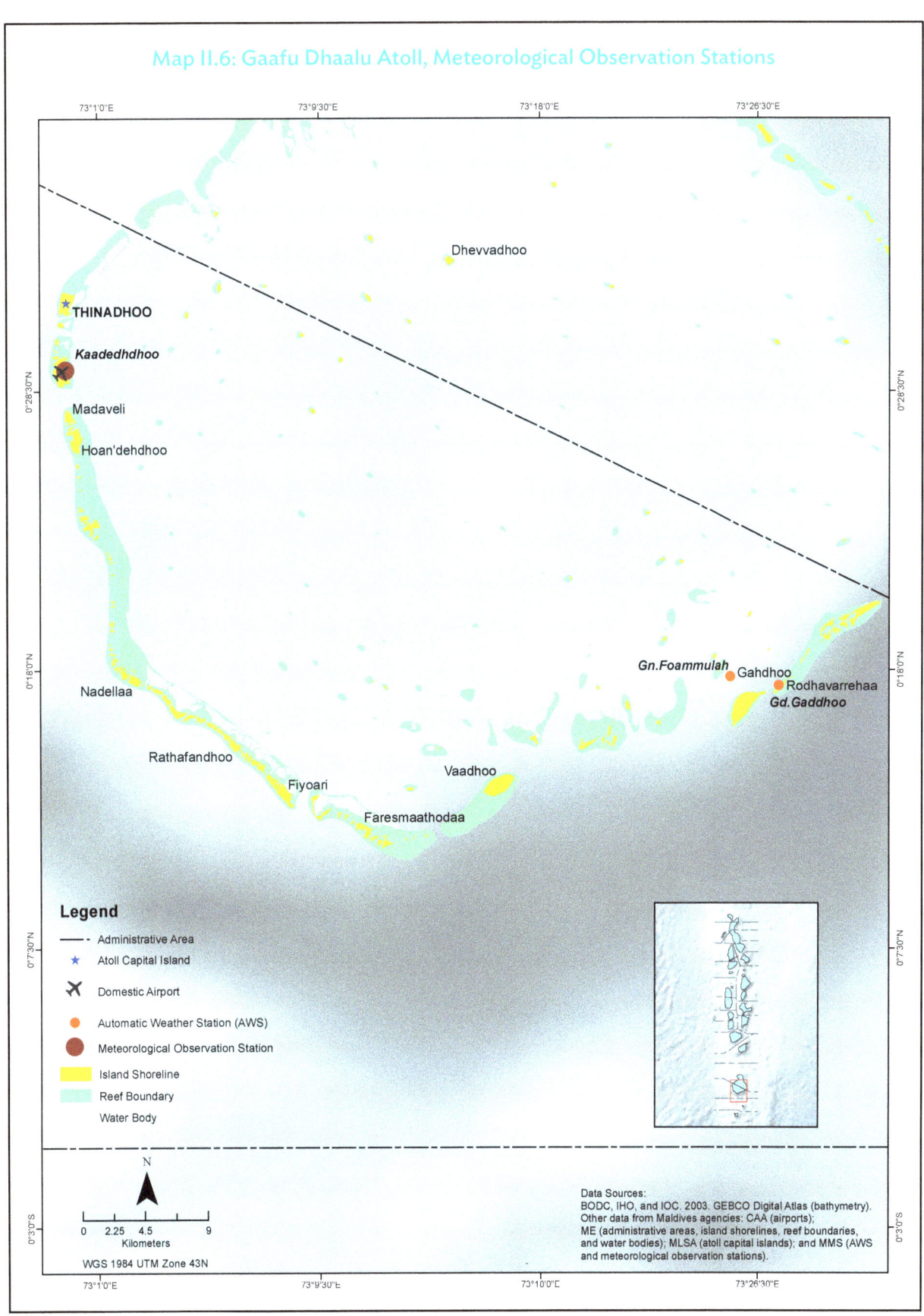

10 Multihazard Risk Atlas of Maldives—Climate and Geophysical Hazards

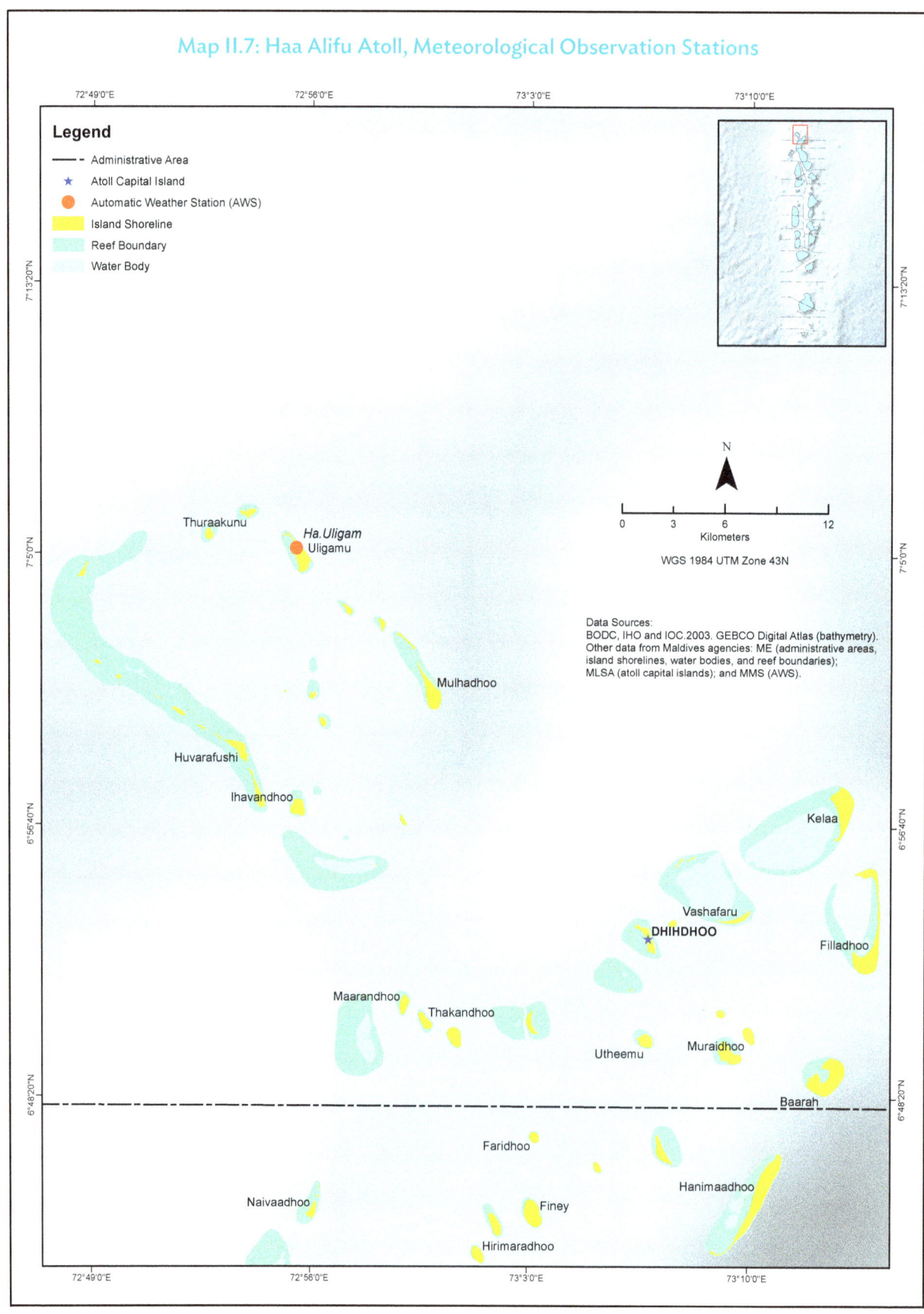

Map II.7: Haa Alifu Atoll, Meteorological Observation Stations

Automatic Weather Stations and Meteorological Observation Stations 11

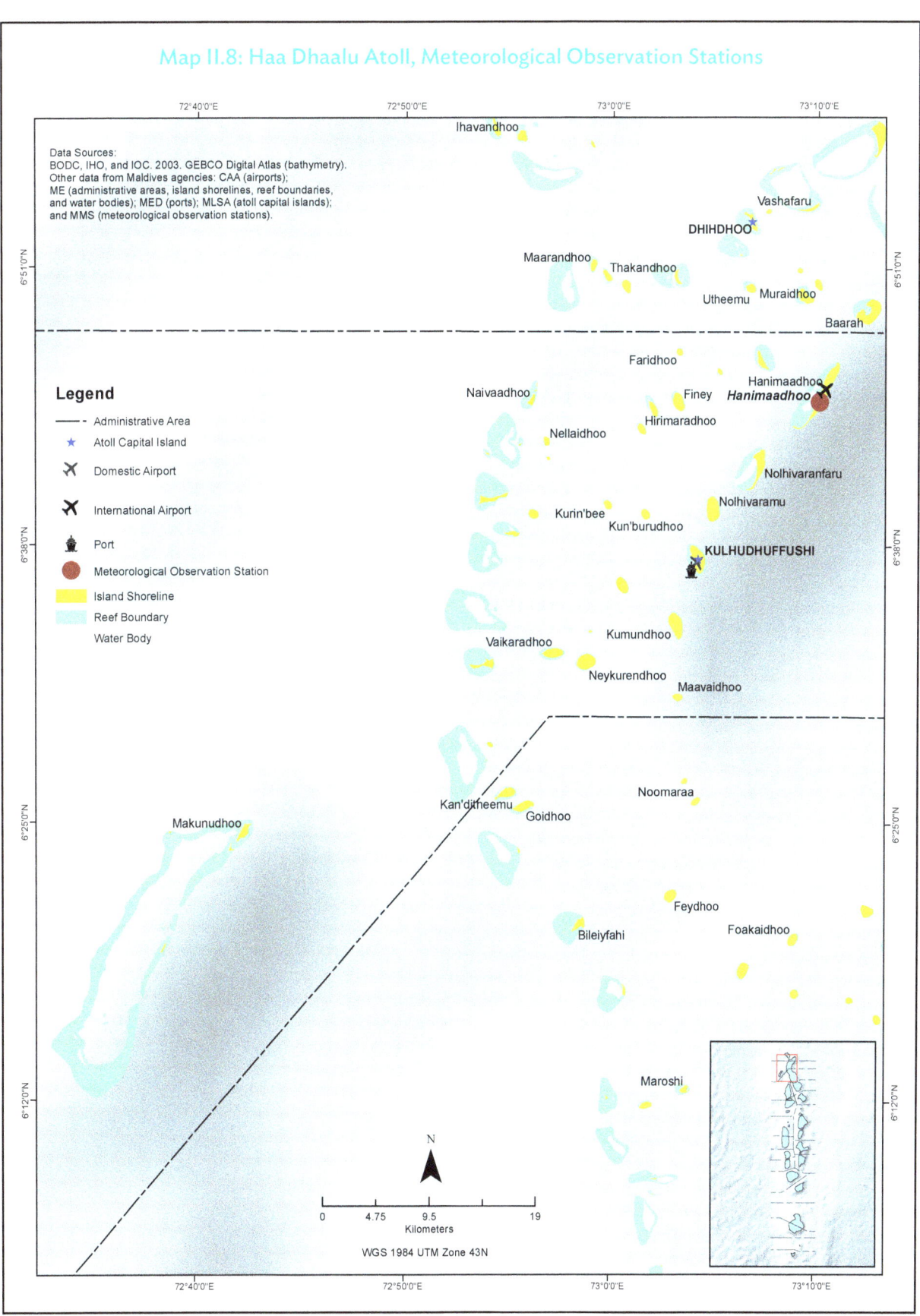

Map II.8: Haa Dhaalu Atoll, Meteorological Observation Stations

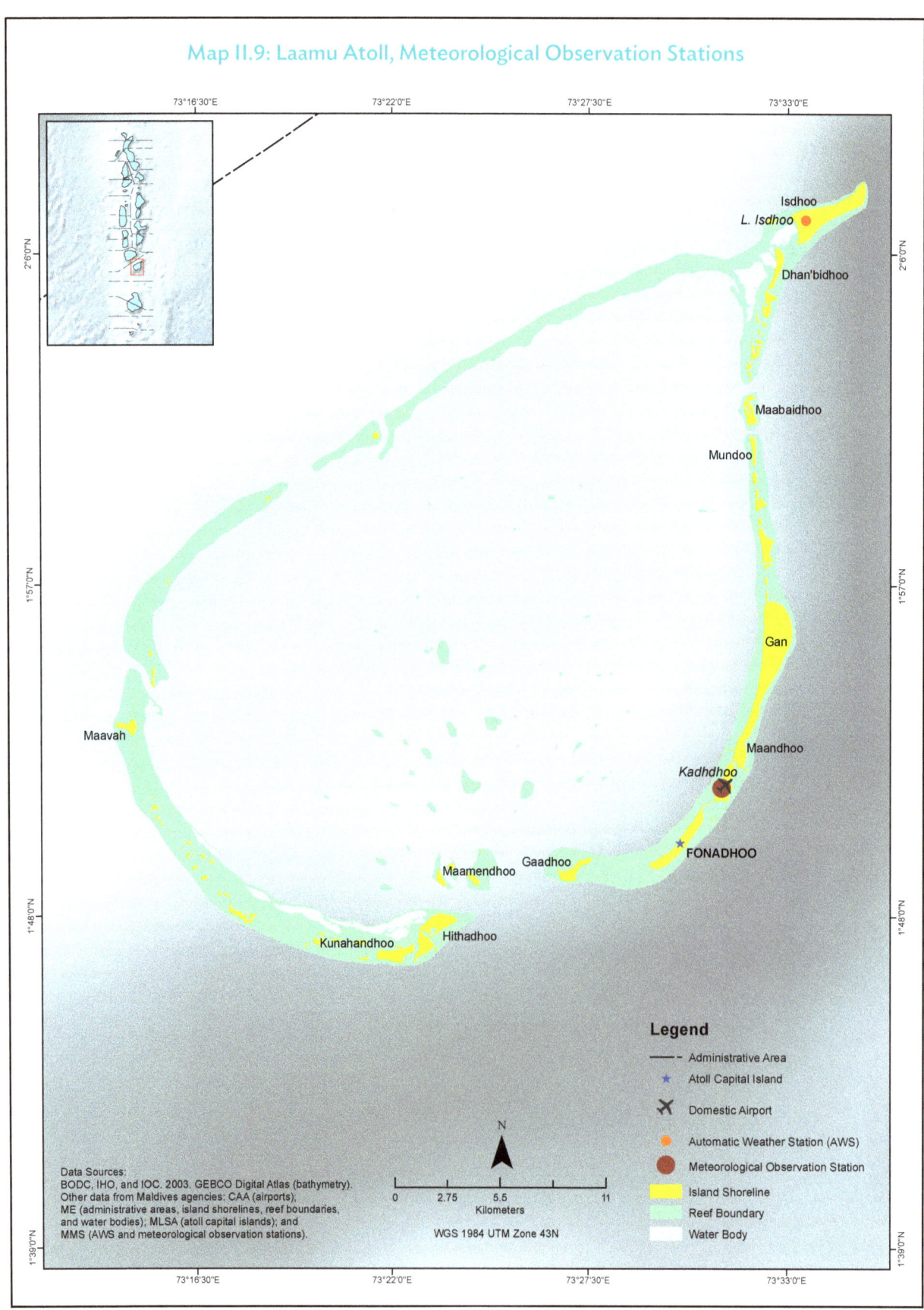

Map II.9: Laamu Atoll, Meteorological Observation Stations

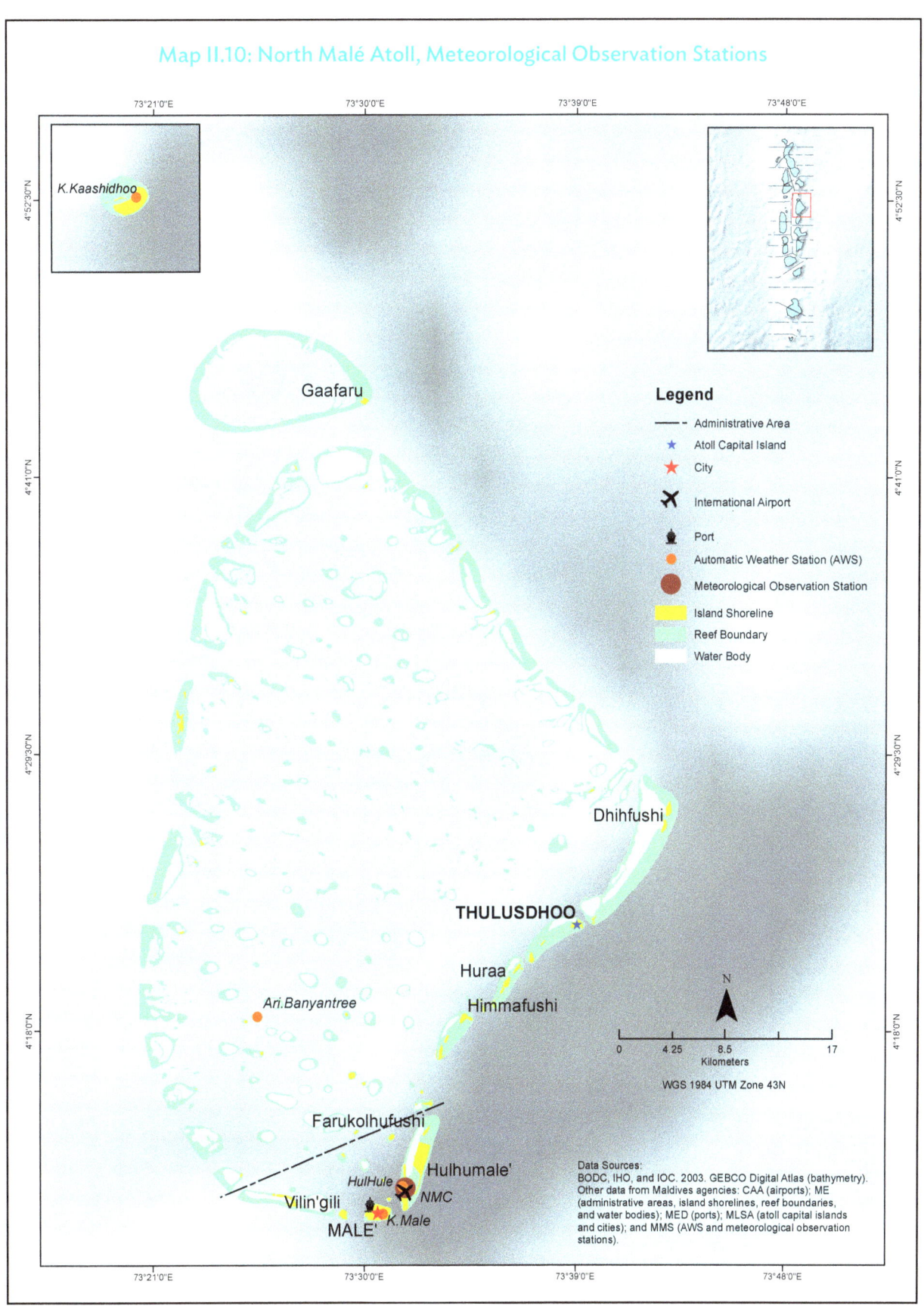

14 Multihazard Risk Atlas of Maldives—Climate and Geophysical Hazards

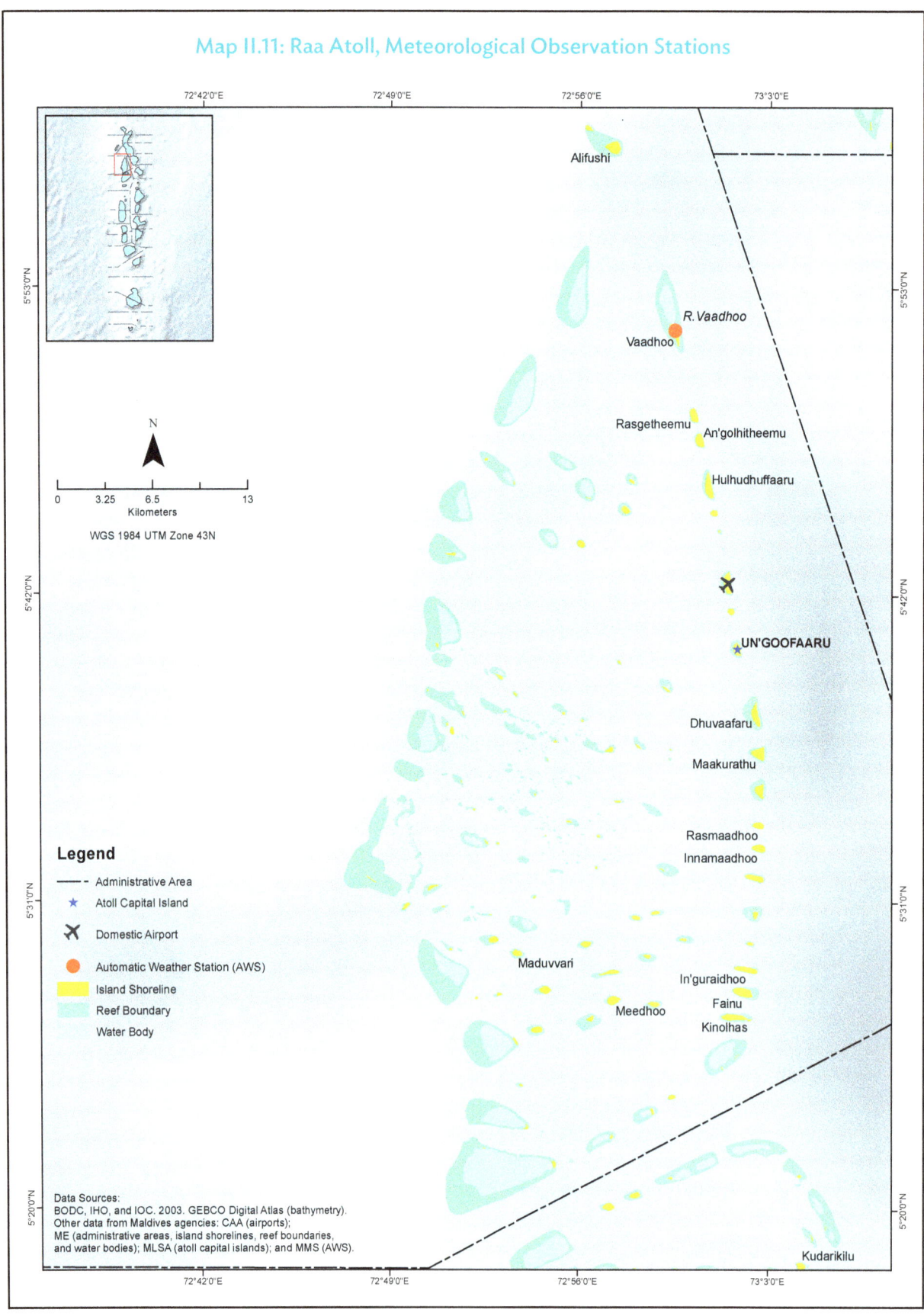

Map II.11: Raa Atoll, Meteorological Observation Stations

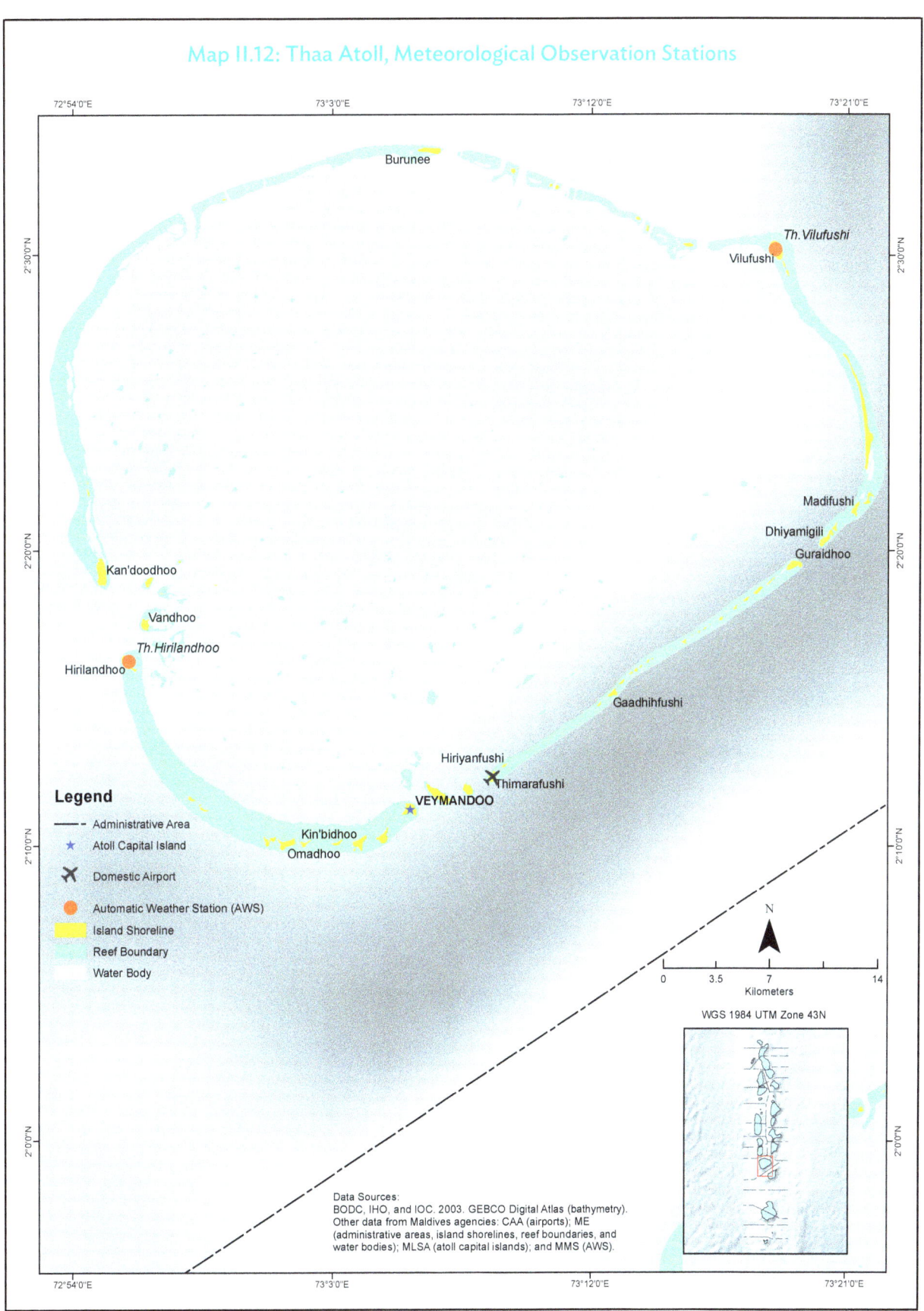

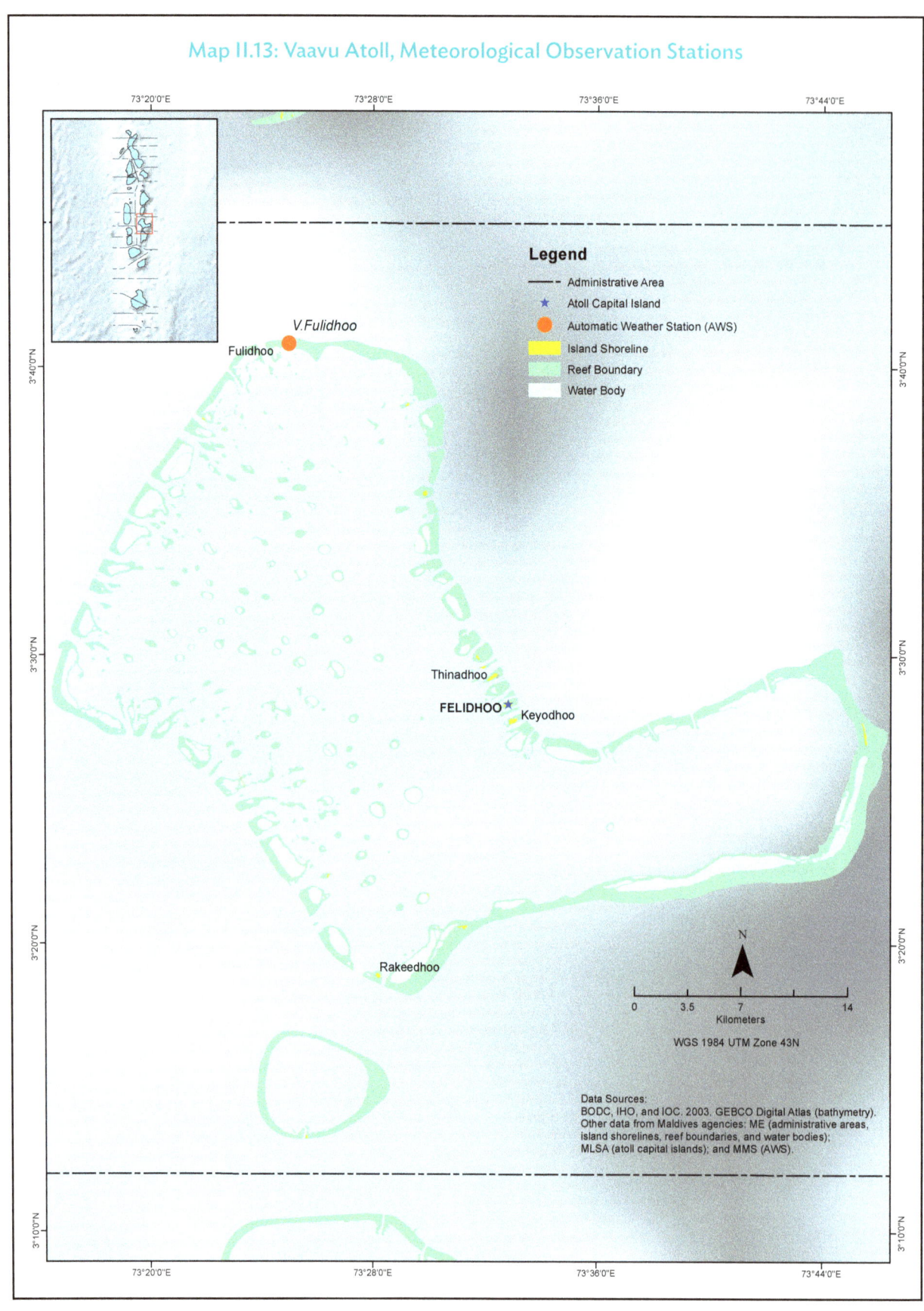

Historical Annual Climate

Daily rainfall and temperature data for the baseline years of 1970–2005 from the Global Historical Climatology Network (GHCN) were averaged and interpolated using thin plate splines for historical annual and seasonal climate mapping.

Rainfall

The resulting annual average rainfall map for 1970–2005 (Map II.14) shows that the least amount of rainfall is received by the atolls in the middle portion of Maldives, from Baa Atoll down to Gnaviyani. These atolls receive an average daily rainfall of 2.1–3.4 millimeters a day (mm/day) and face a greater challenge in sourcing their water supply. On the contrary, rainfall is highest in the northern atolls (Haa Alifu and Haa Dhaalu). These atolls receive an average of 4.6–5.8 mm/day. The rest of the atolls receive an average of 3.4–4.6 mm/day.

Temperature

The annual average temperature map for 1970–2005 (Map II.15) illustrates a slight difference in average annual temperature across Maldives. Higher latitudes have slightly higher average annual temperature (28.0°C–28.4°C) compared with the southern atolls. Southern atolls—from Meemu Atoll and Dhaalu Atoll down to Addu Atoll—have an average annual temperature of 27.7°C–28.0°C.

Sunset view from the Malé fishport. Fishing is the dominant economic activity and largest sector in Maldives, followed by tourism. Trends in these industries can be affected by the country's tropical climate, hot and humid all year-round, with rainfall influenced by monsoons (photo by Roberta Gerpacio).

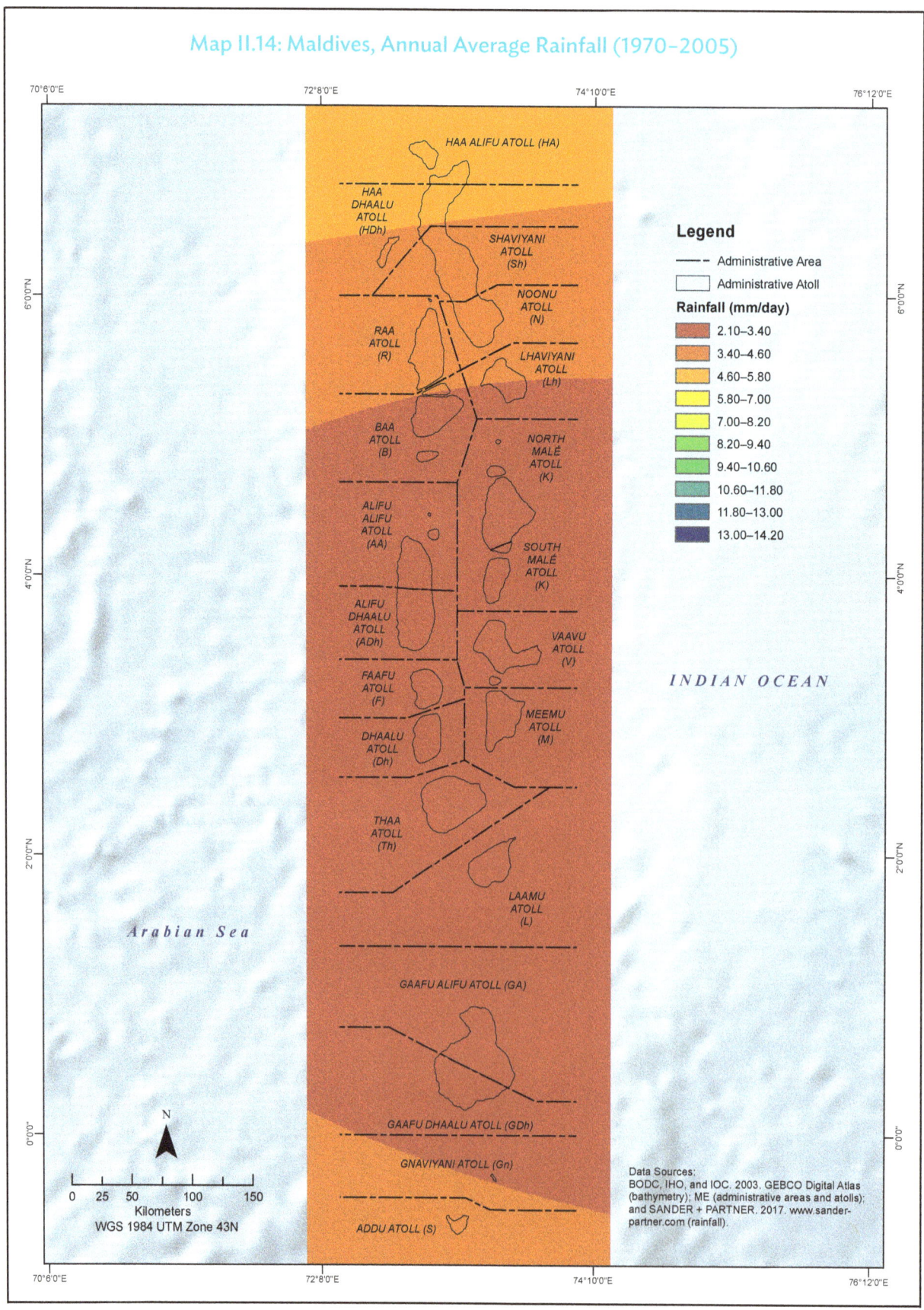

Map II.14: Maldives, Annual Average Rainfall (1970–2005)

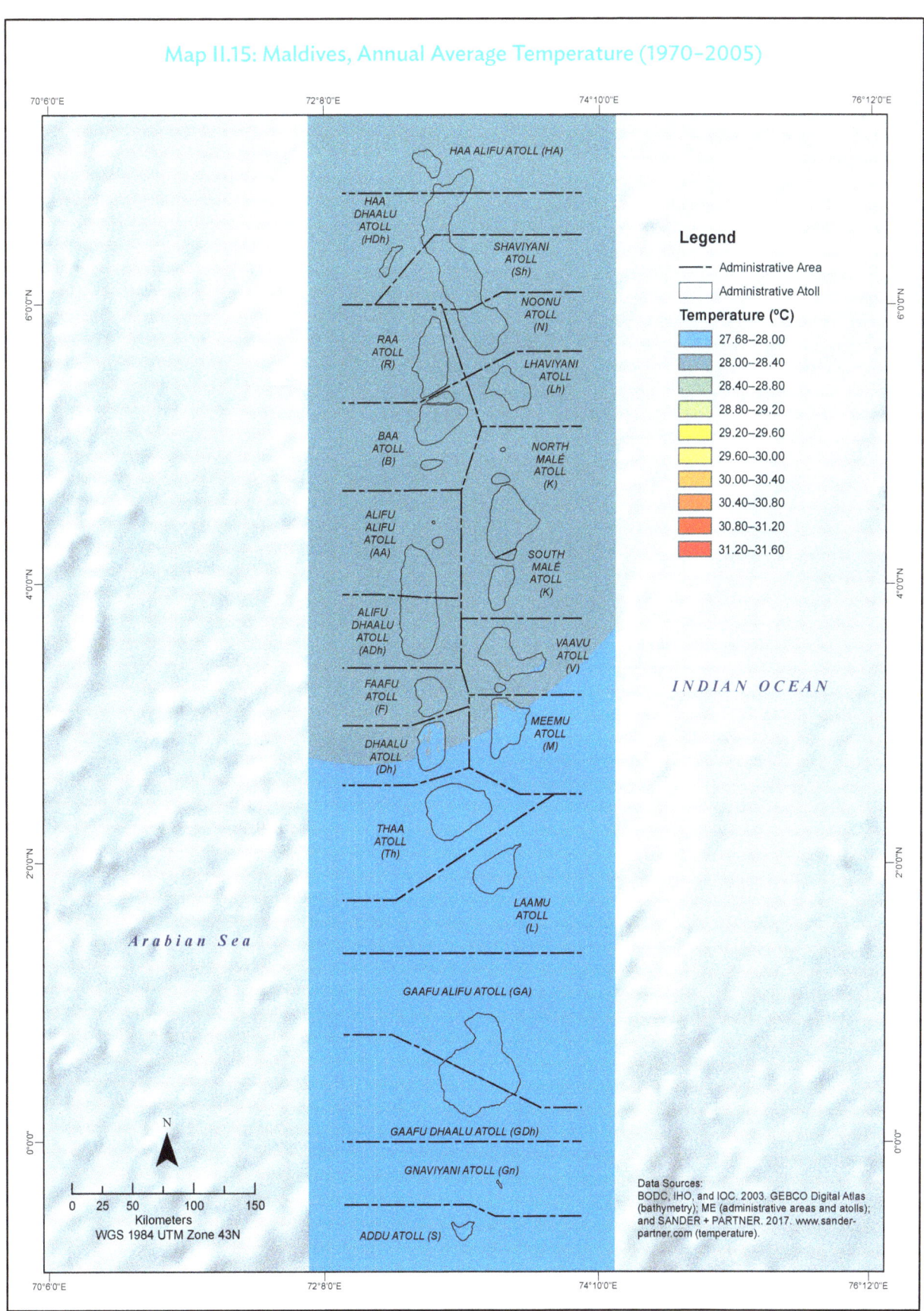

Map II.15: Maldives, Annual Average Temperature (1970–2005)

Historical Seasonal Climate

For seasonal climate, the same dataset from GHCN was used. These maps are presented by season: December, January, February (DJF); March, April, May (MAM); June, July, August (JJA); and September, October, November (SON). The maps show rainfall and temperature variation across time and space.

Shoreline in a resort island. Maldives has developed numerous resort islands to serve the growing tourism industry, an important catalyst in the country's overall economic growth. The industry, however, can also be vulnerable to the impacts of changing climate (photo by Roberta Gerpacio).

Seasonal Average Rainfall (1970–2005)

21

The rainfall distribution pattern across Maldives varies across months and latitude. Greater rainfall is experienced in the south during DJF. During MAM, Maldives is generally dry. The climate is governed by monsoons. MAM covers the transition of a drier northeast monsoon to a wetter southwest monsoon. Rainfall starts to increase in JJA, during which rainfall is greater in the northern atolls and the country experiences rainfall from the southwest monsoon. The southwest monsoon continues to bring rainfall until October. The wettest months are SON—northern and southern atolls experience greater rainfall during these months.

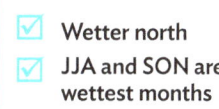
☑ Wetter north
☑ JJA and SON are wettest months

Map II.16: Maldives, Seasonal Average Rainfall (1970–2005)

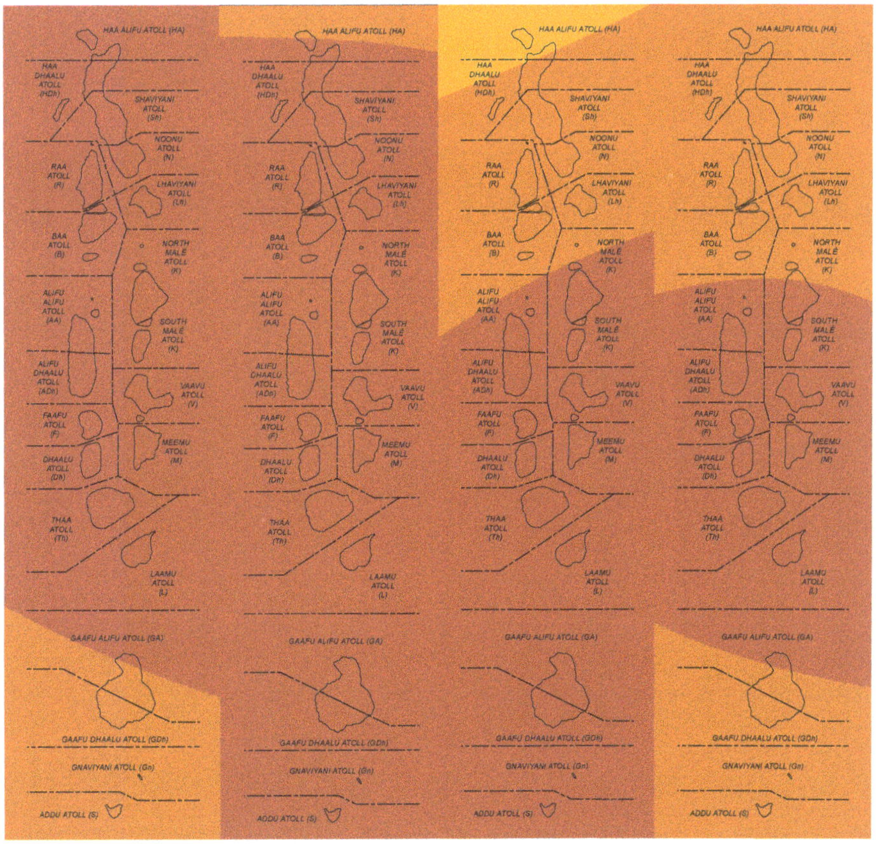

Legend
— — Administrative Area
☐ Administrative Atoll

Rainfall (mm/day)
- 1.30–4.00
- 4.00–8.00
- 8.00–12.00
- 12.00–16.00
- 16.00–20.00
- 20.00–24.00
- 24.00–28.00
- 28.00–32.00
- 32.00–36.00
- 36.00–44.00

Data Sources:
BODC, IHO, and IOC. 2003.
GEBCO Digital Atlas (bathymetry);
ME (administrative areas and atoll);
and SANDER + PARTNER. 2017.
www.sander-partner.com (rainfall).

DJF   MAM   JJA   SON

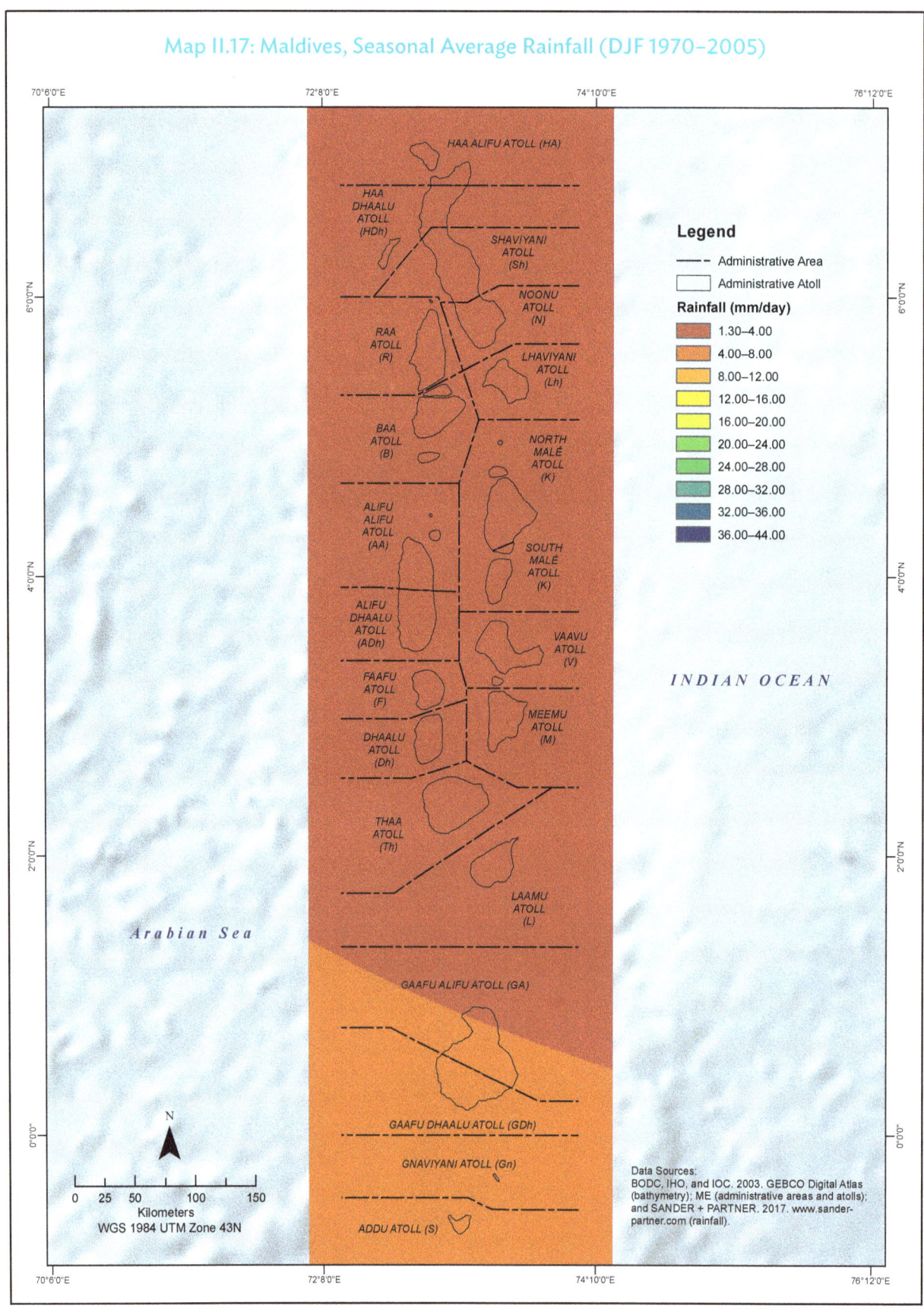

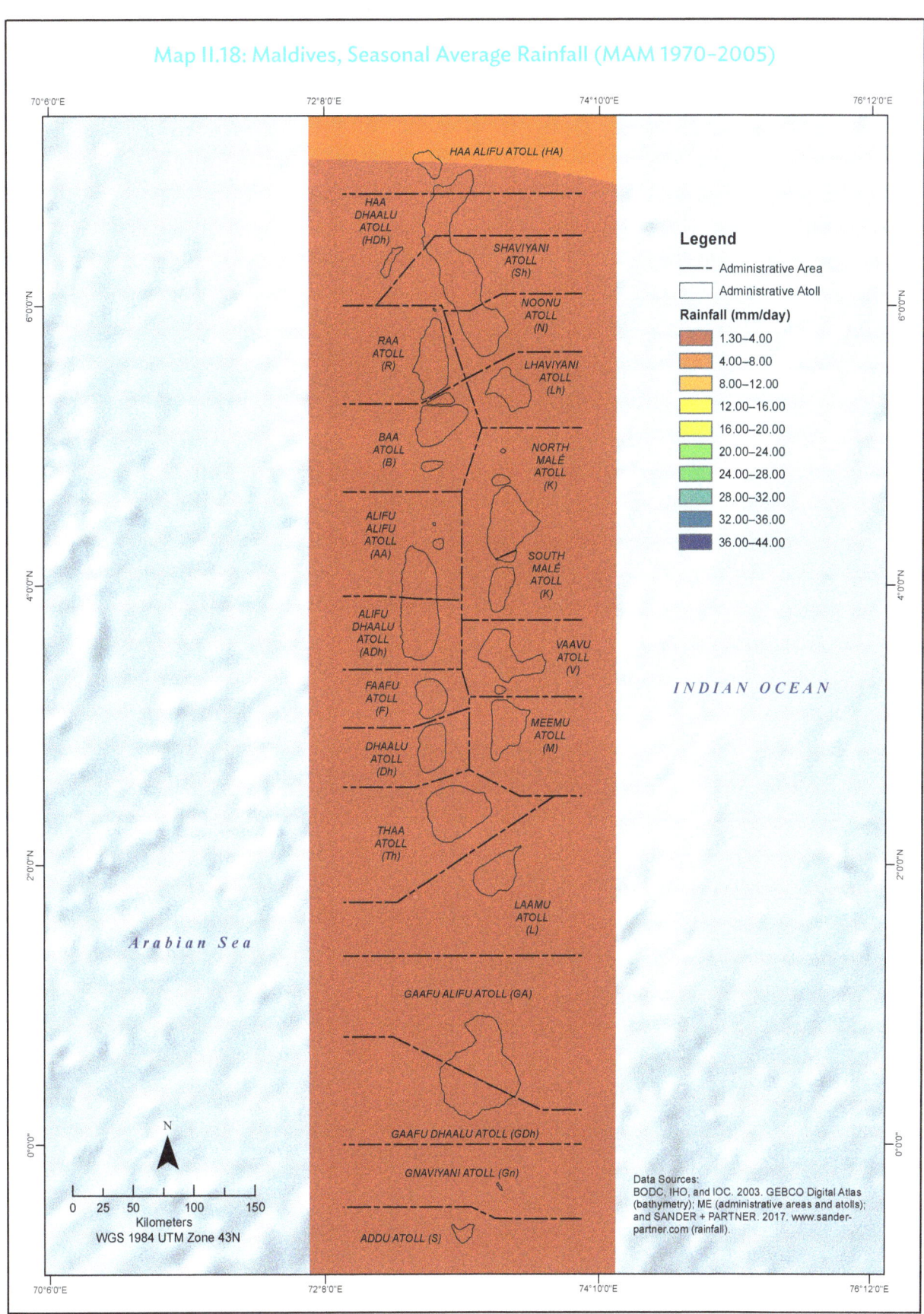

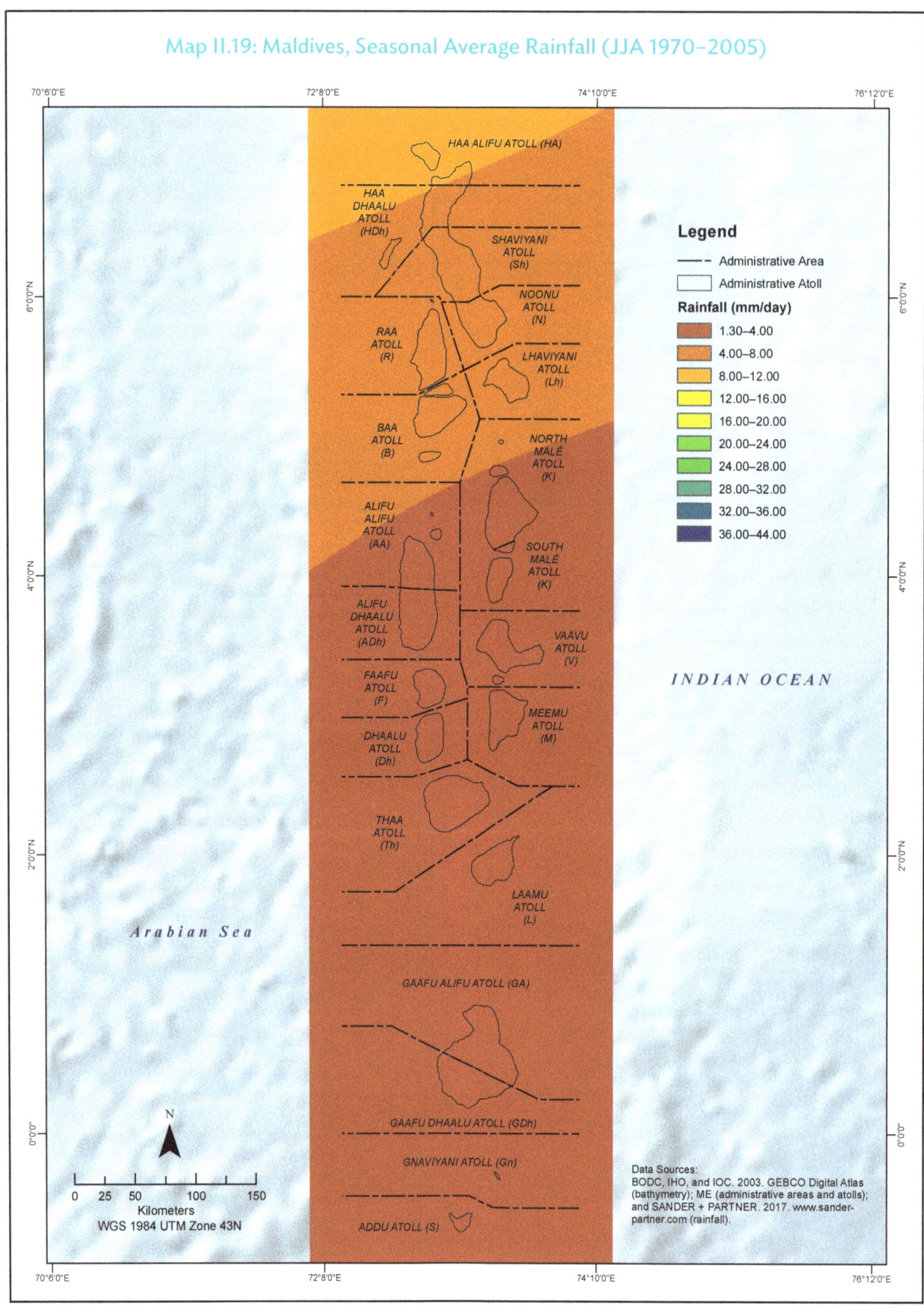

Map II.19: Maldives, Seasonal Average Rainfall (JJA 1970–2005)

Historical Seasonal Climate

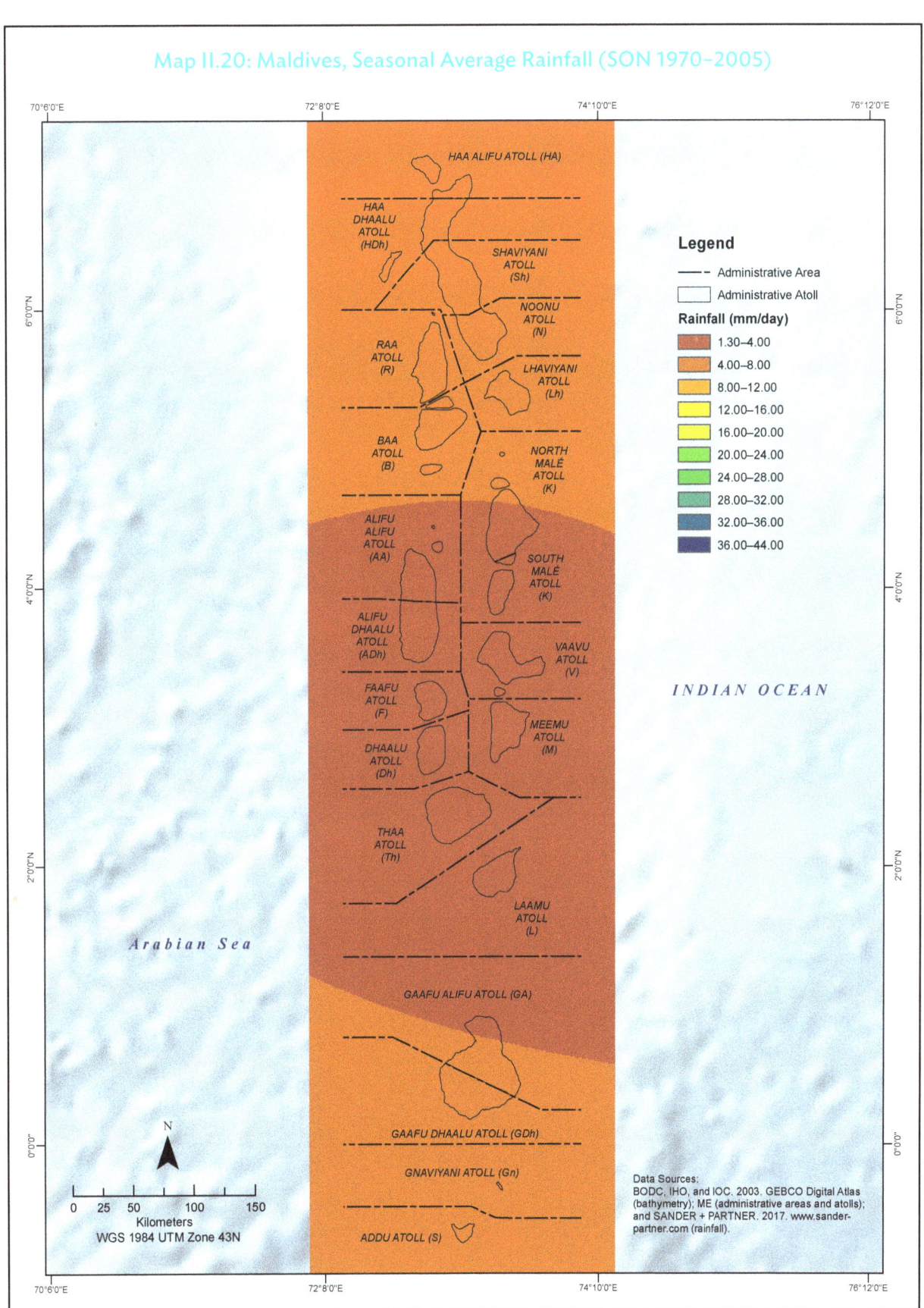

Map II.20: Maldives, Seasonal Average Rainfall (SON 1970–2005)

Seasonal Average Temperature (1970–2005)

- ✓ Warmer north
- ✓ MAM are the warmest months

Maldives is coldest during DJF and SON. The average temperature, ranging 27.40°C–27.95°C, is consistent across the country. The temperature (27.95°C–30.15°C) during MAM increases. MAM are the warmest months especially in the northern atolls. During JJA, the temperature decreases to 27.40°C–27.95 °C in the southern atolls and 27.95°C–28.50 °C in the northern and central atolls.

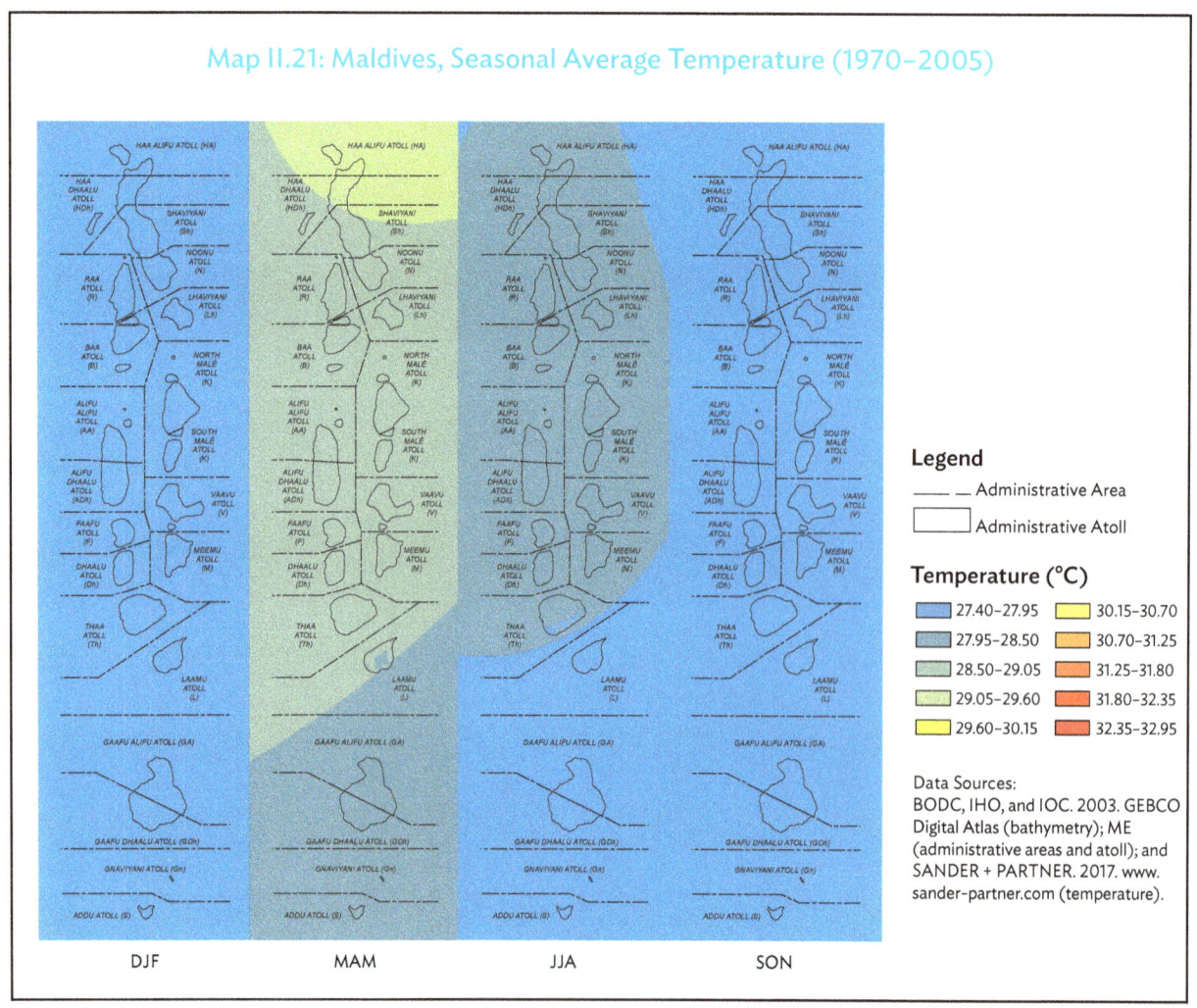

Map II.21: Maldives, Seasonal Average Temperature (1970–2005)

Historical Seasonal Climate 27

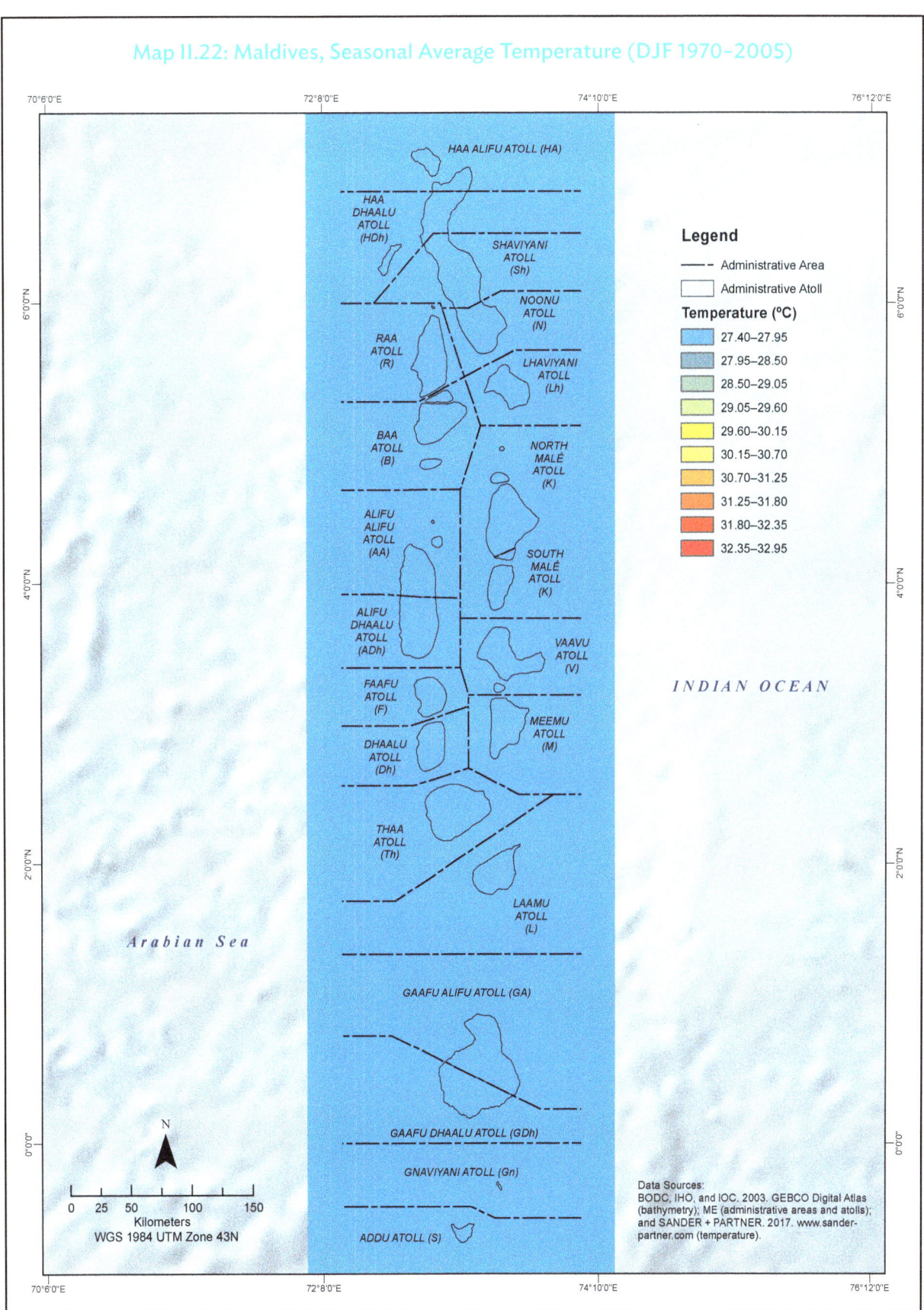

Map II.22: Maldives, Seasonal Average Temperature (DJF 1970–2005)

28 Multihazard Risk Atlas of Maldives—Climate and Geophysical Hazards

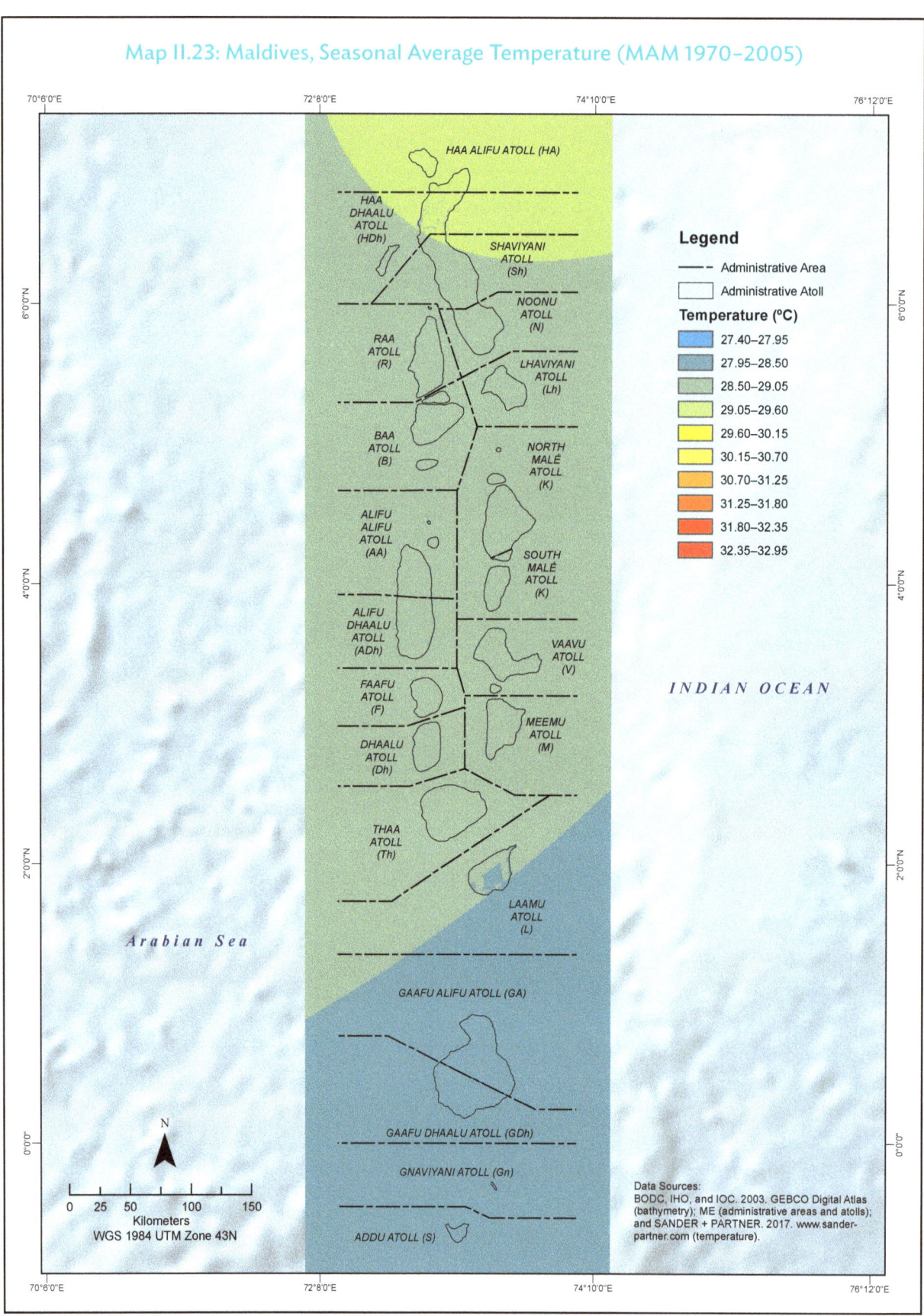

Map II.23: Maldives, Seasonal Average Temperature (MAM 1970–2005)

Historical Seasonal Climate

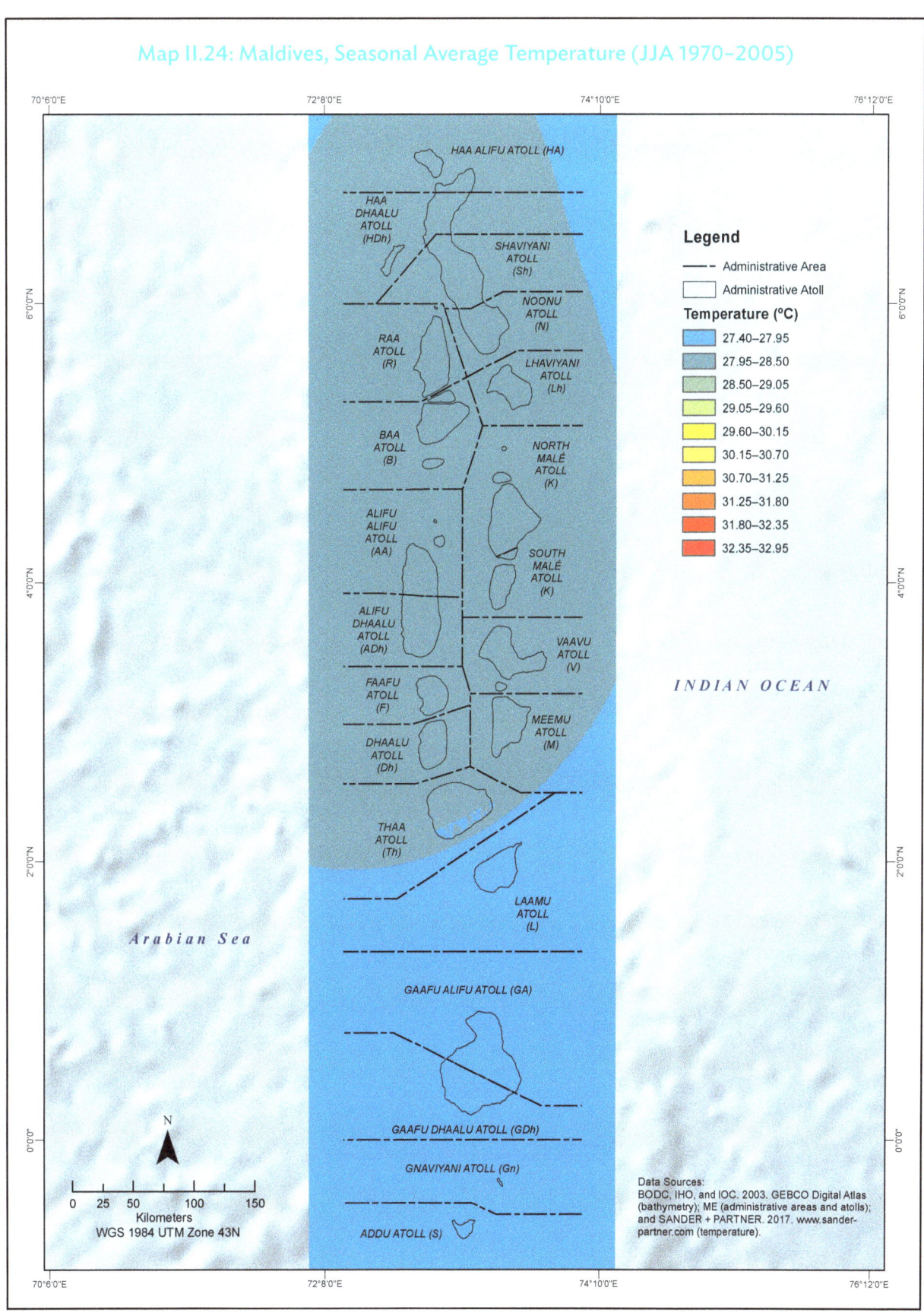

Map II.24: Maldives, Seasonal Average Temperature (JJA 1970–2005)

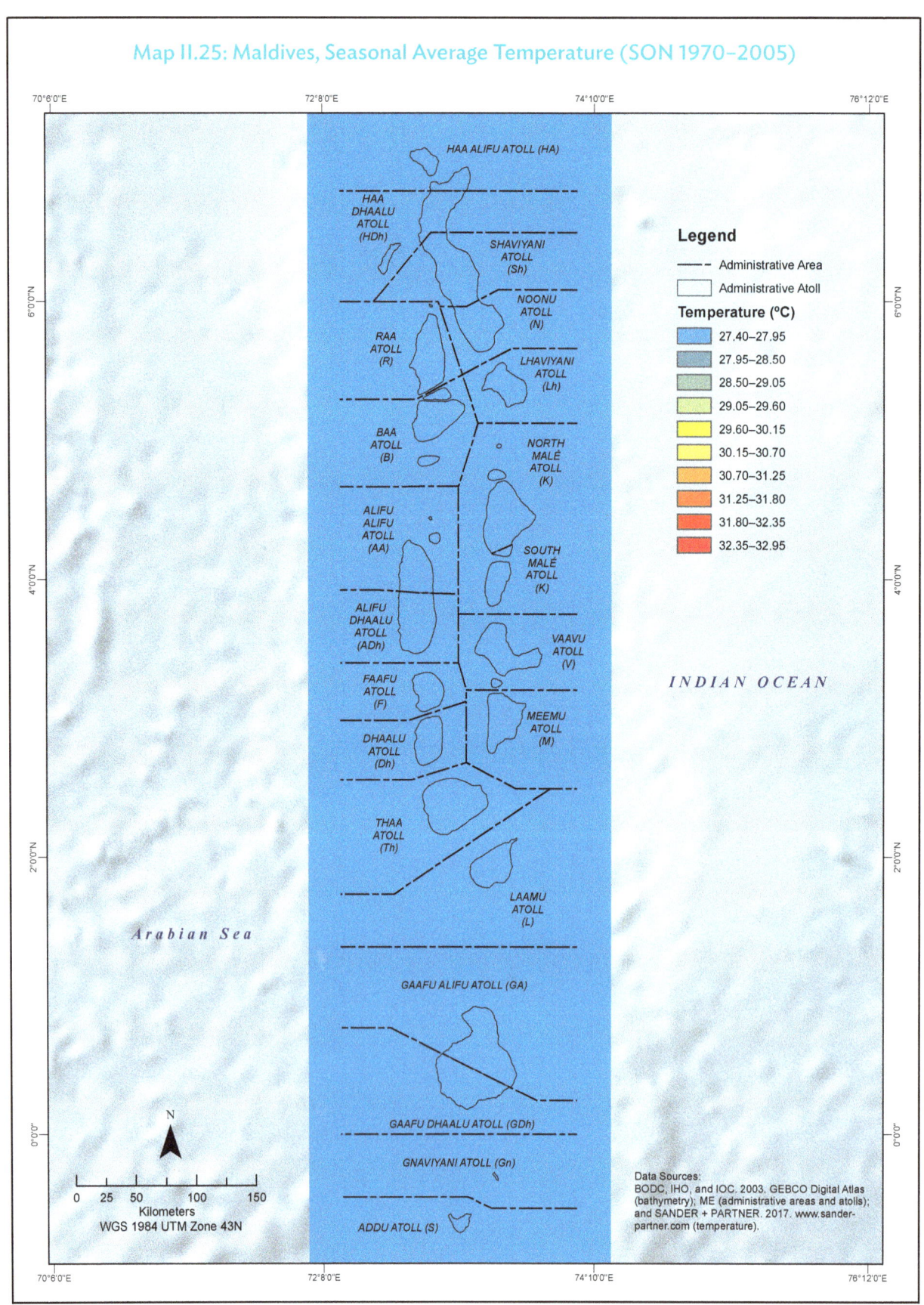

Future Climate

Throughout the formation of our planet, the climate has changed and will continue to change. Together with geological processes, climate change continues to shape Maldives. While the historical climate of the country sustained life, future atmospheric conditions may pose several threats to life in these tropical islands.

The same data set from GHCN was downscaled to the regional climate models to have a finer resolution of 50 kilometers. The model followed the RCP based on the Intergovernmental Panel on Climate Change's Fifth Assessment Report (IPCC 2014). Two RCPs (4.5 and 8.5) were selected. RCP 4.5 represents a moderate greenhouse gas emission scenario, while RCP 8.5 represents a high greenhouse gas emission scenario with no mitigation efforts. Results were further downscaled to 900 meters.

The downscaled projected climate shows a warmer and wetter Maldives in the future. The succeeding maps will illustrate the projected rainfall and temperature per decade to give a clearer picture of the changing climate.

Lone tree. A leafless tree standing in a rocky shore in Maldives. Future changes in climate will have a tremendous impact on the life cycle of trees and other forms of living organisms (photo by Ahmed Shareef).

Average Annual Rainfall Projection (RCP 4.5)

- Wetter north
- Northern atolls will be wetter in 2030s and 2040s.

According to the projected rainfall based on RCP 4.5 (moderate greenhouse gas emissions), Haa Alifu and Shaviyani atolls as well as Baa and Lhaviyani atolls will experience an increase in average annual rainfall rate in the coming decades. The maps show a slight decrease in the area with the 1.3-4 mm/day rainfall rate, indicating a general increase in rainfall rate through the decades. A southward trend in increased rainfall can also be observed in the projected average annual rainfall across the decades.

Map II.26: Maldives, Annual Average Rainfall Projection (RCP 4.5)

2011–2020 2021–2030 2031–2040 2041–2050

Legend
- Administrative Area
- Administrative Atoll

Rainfall (mm/day)
- 1.30–4.00
- 4.00–8.00
- 8.00–12.00
- 12.00–16.00
- 16.00–20.00
- 20.00–24.00
- 24.00–28.00
- 28.00–32.00
- 32.00–36.00
- 36.00–44.00

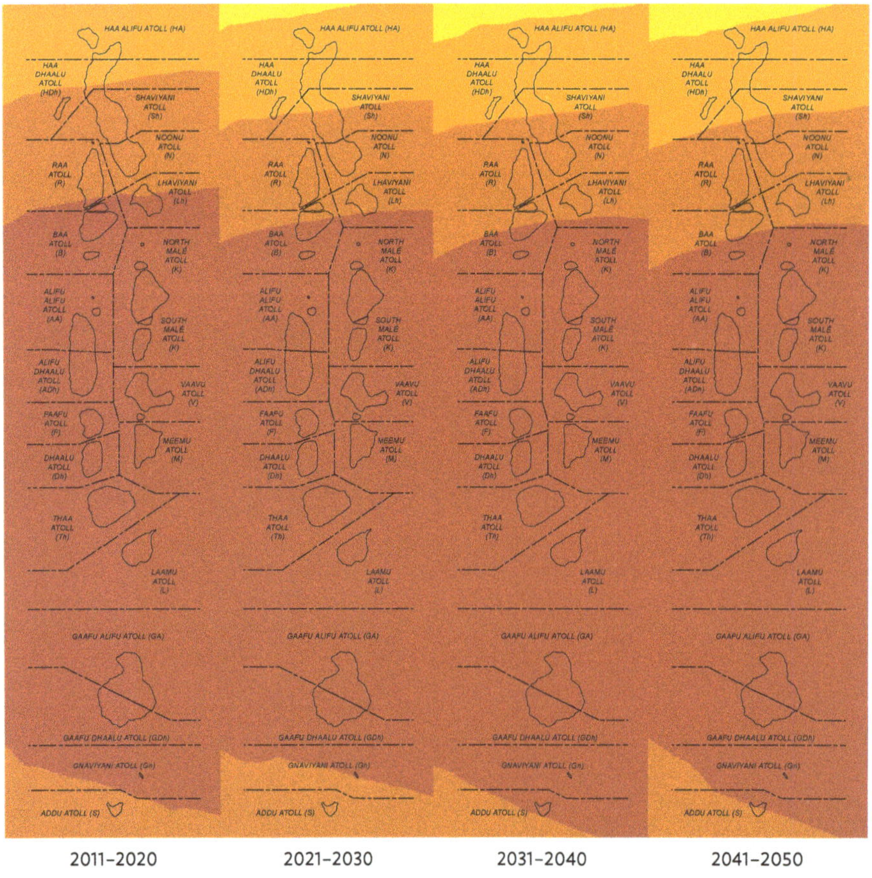

Data Sources:
BODC, IHO, and IOC. 2003. GEBCO Digital Atlas (bathymetry); ME (administrative areas and atoll); and SANDER + PARTNER. 2017. www.sander-partner.com (rainfall).

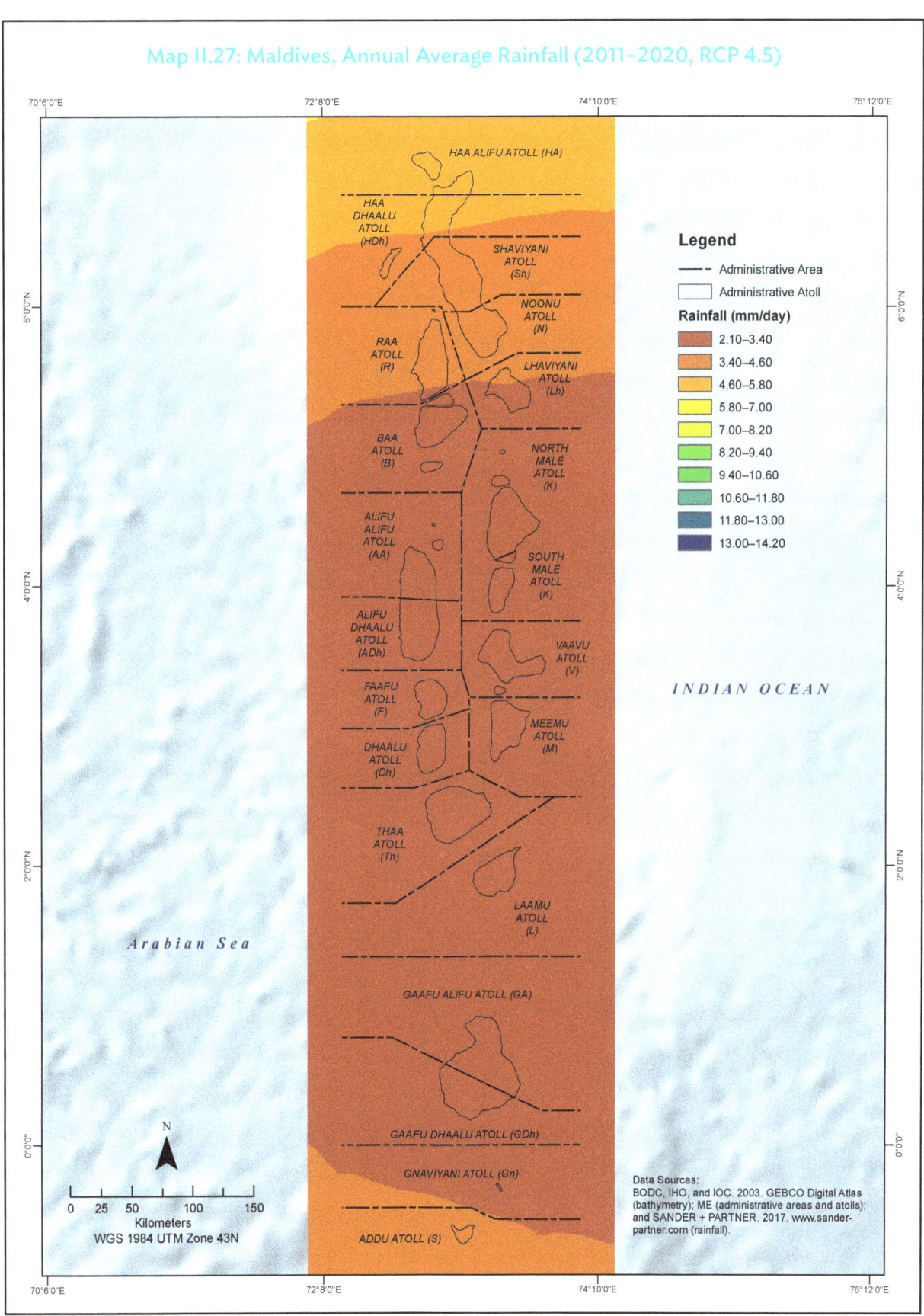

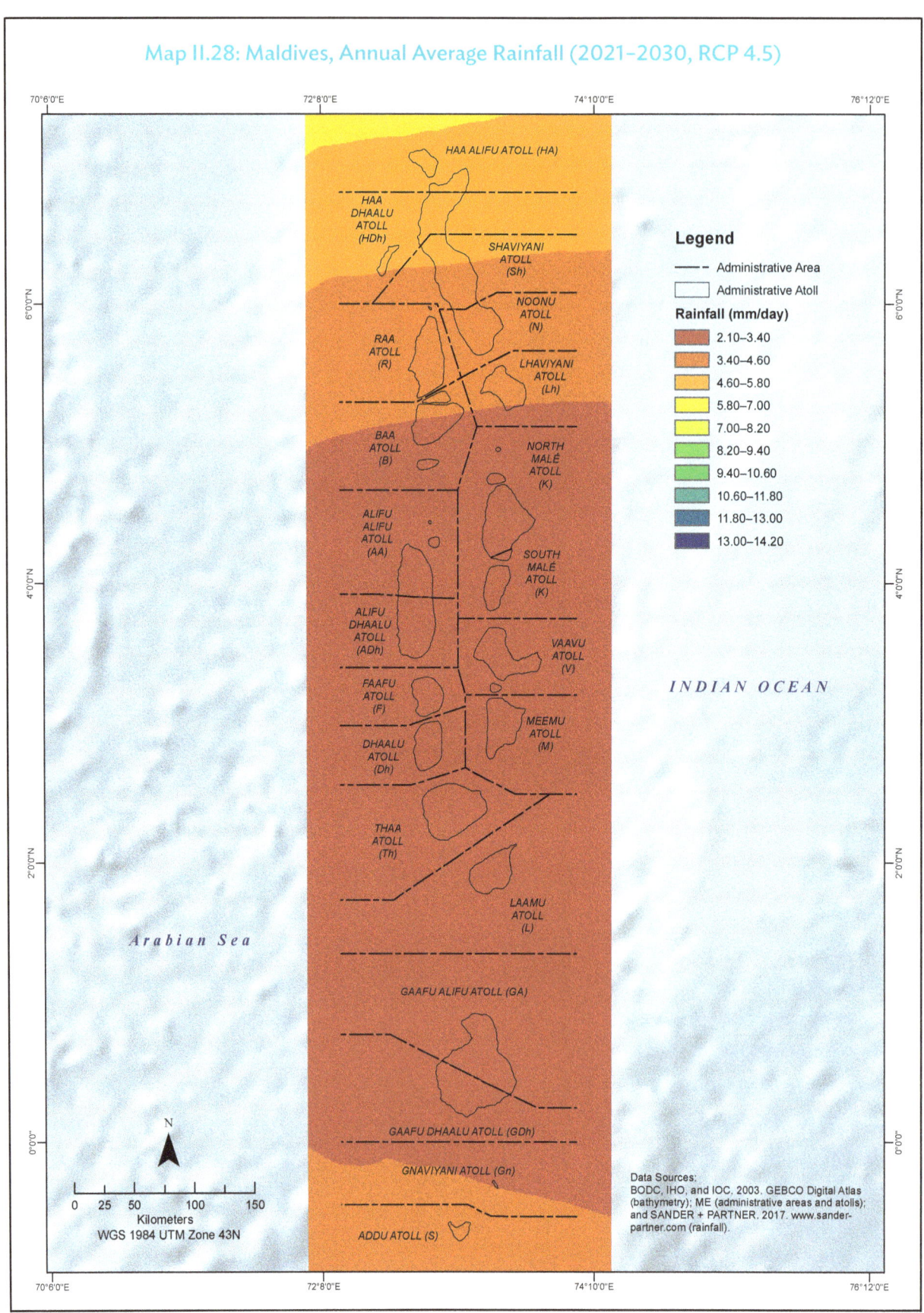

Map II.28: Maldives, Annual Average Rainfall (2021–2030, RCP 4.5)

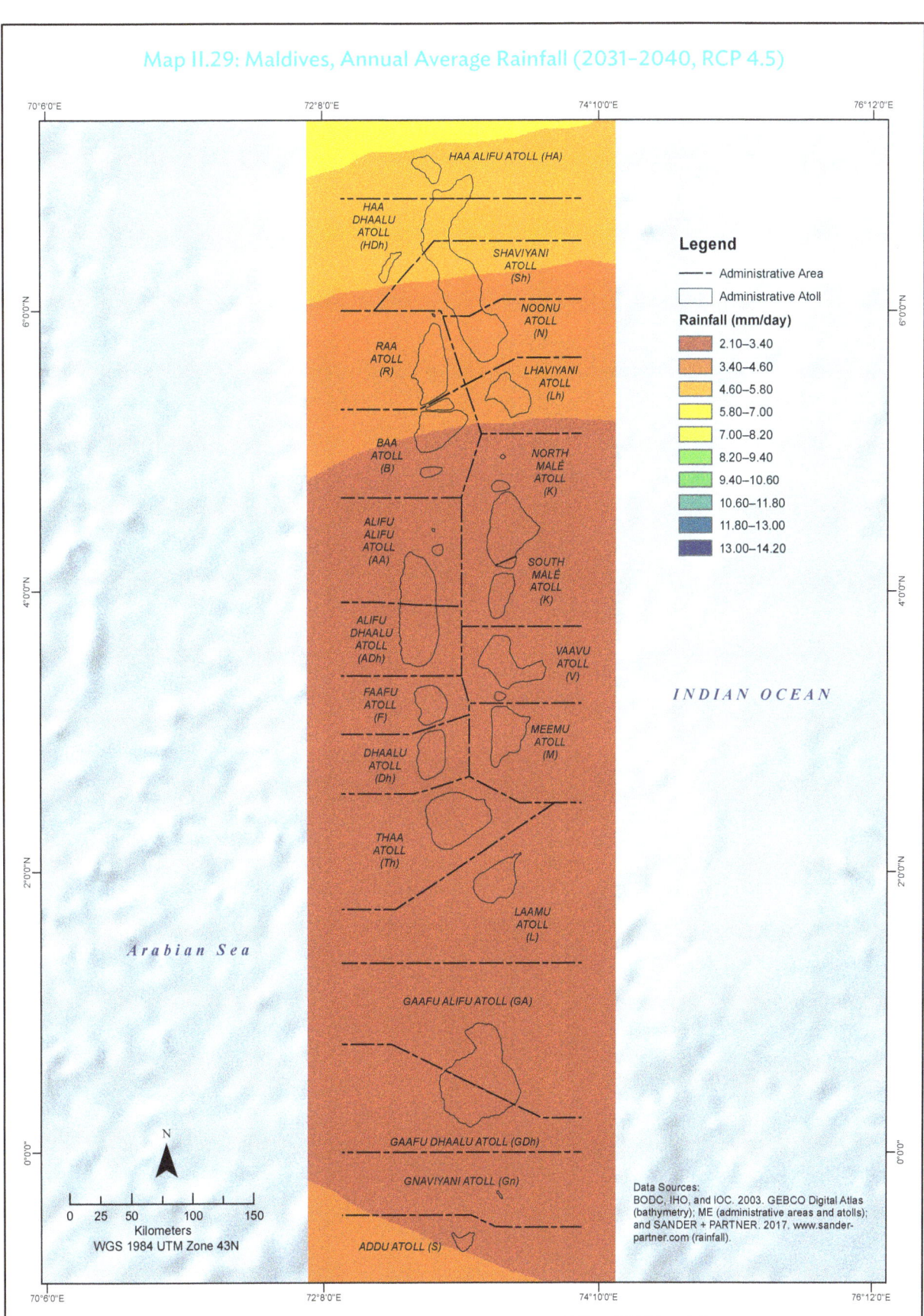

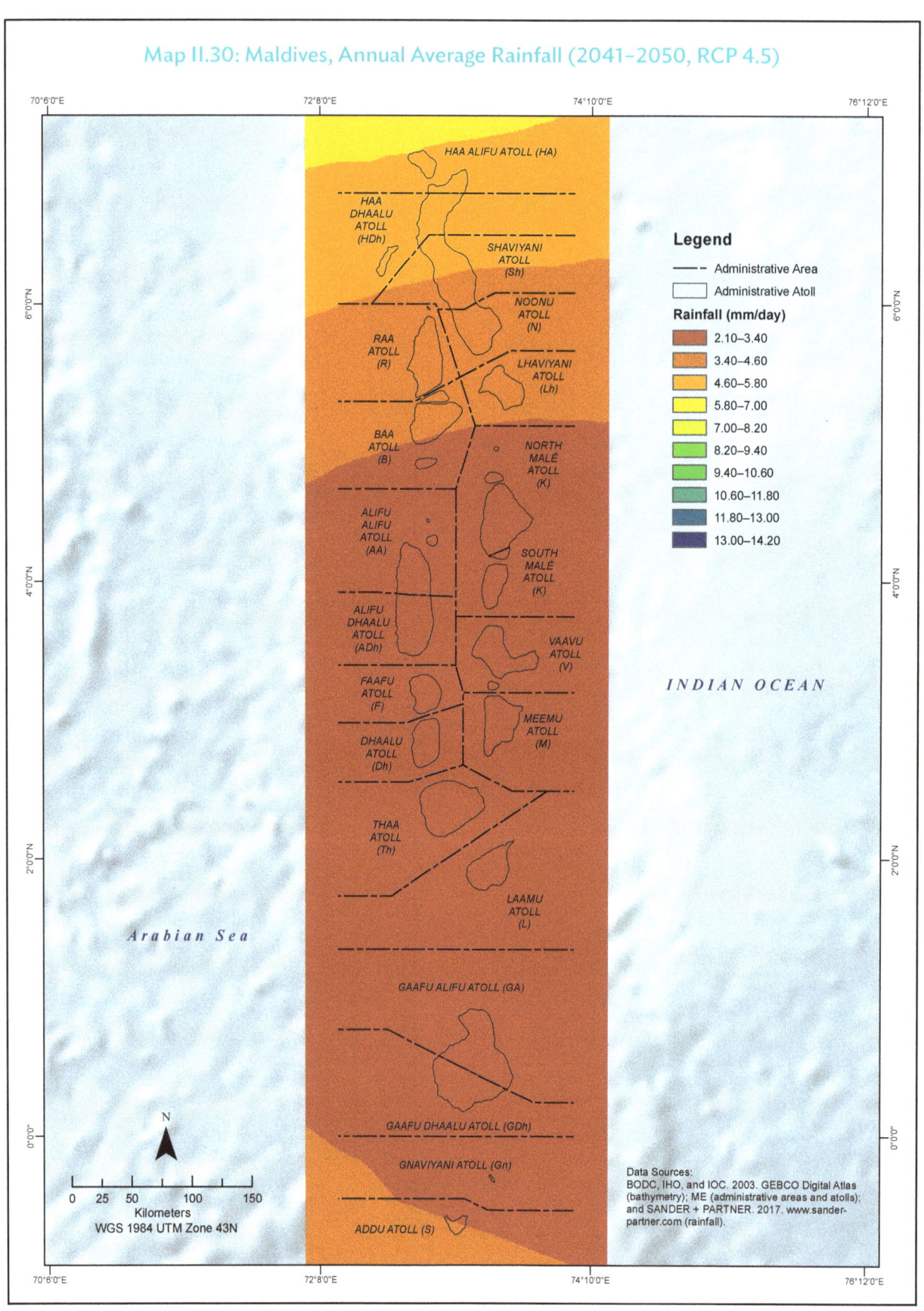
Map II.30: Maldives, Annual Average Rainfall (2041–2050, RCP 4.5)

Average Seasonal Rainfall Projection (DJF, RCP 4.5)

The northeast monsoon prevails during DJF, keeping rainfall in the northern and middle portion of Maldives low (1.3–4 mm/day). These months bring rainfall (4–8 mm/day) to the southern atolls. The southern atolls will experience wetter 2030s and 2040s.

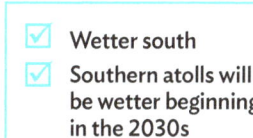

- ✓ Wetter south
- ✓ Southern atolls will be wetter beginning in the 2030s

Map II.31: Maldives, Average Seasonal Rainfall Projection (DJF, RCP 4.5)

2011–2020 2021–2030 2031–2040 2041–2050

Legend
— Administrative Area
☐ Administrative Atoll

Rainfall (mm/day)
- 1.30–4.00
- 4.00–8.00
- 8.00–12.00
- 12.00–16.00
- 16.00–20.00
- 20.00–24.00
- 24.00–28.00
- 28.00–32.00
- 32.00–36.00
- 36.00–44.00

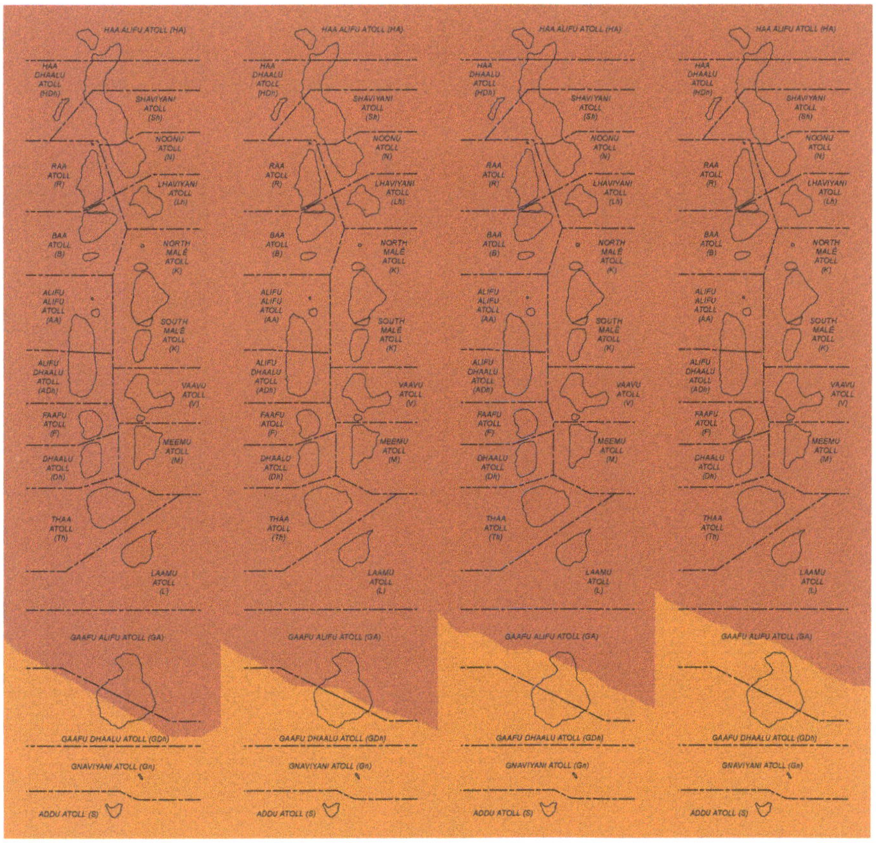

Data Sources:
BODC, IHO, and IOC. 2003.
GEBCO Digital Atlas (bathymetry);
ME (administrative areas and atoll);
and SANDER + PARTNER. 2017.
www.sander-partner.com (rainfall).

38 Multihazard Risk Atlas of Maldives—Climate and Geophysical Hazards

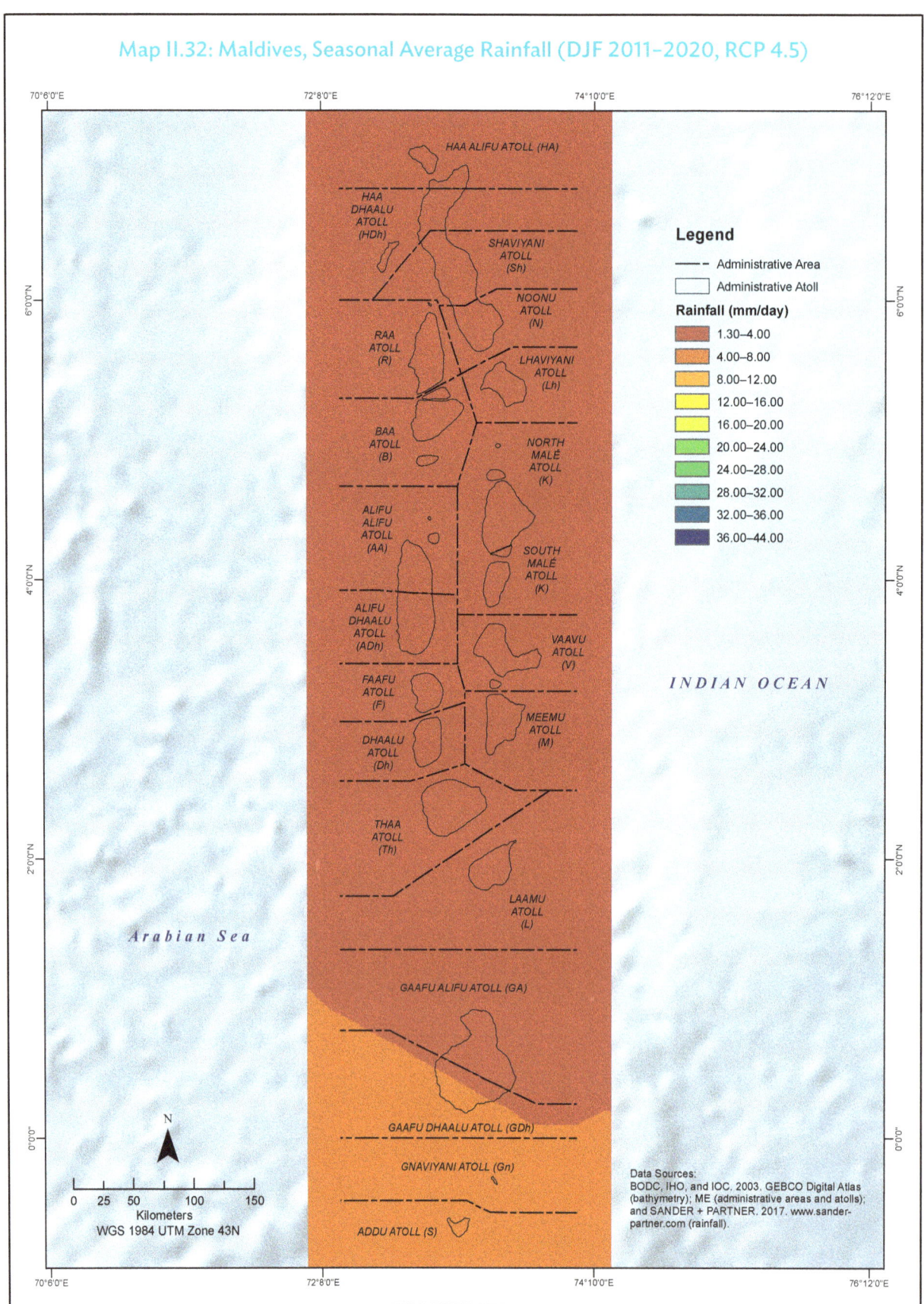

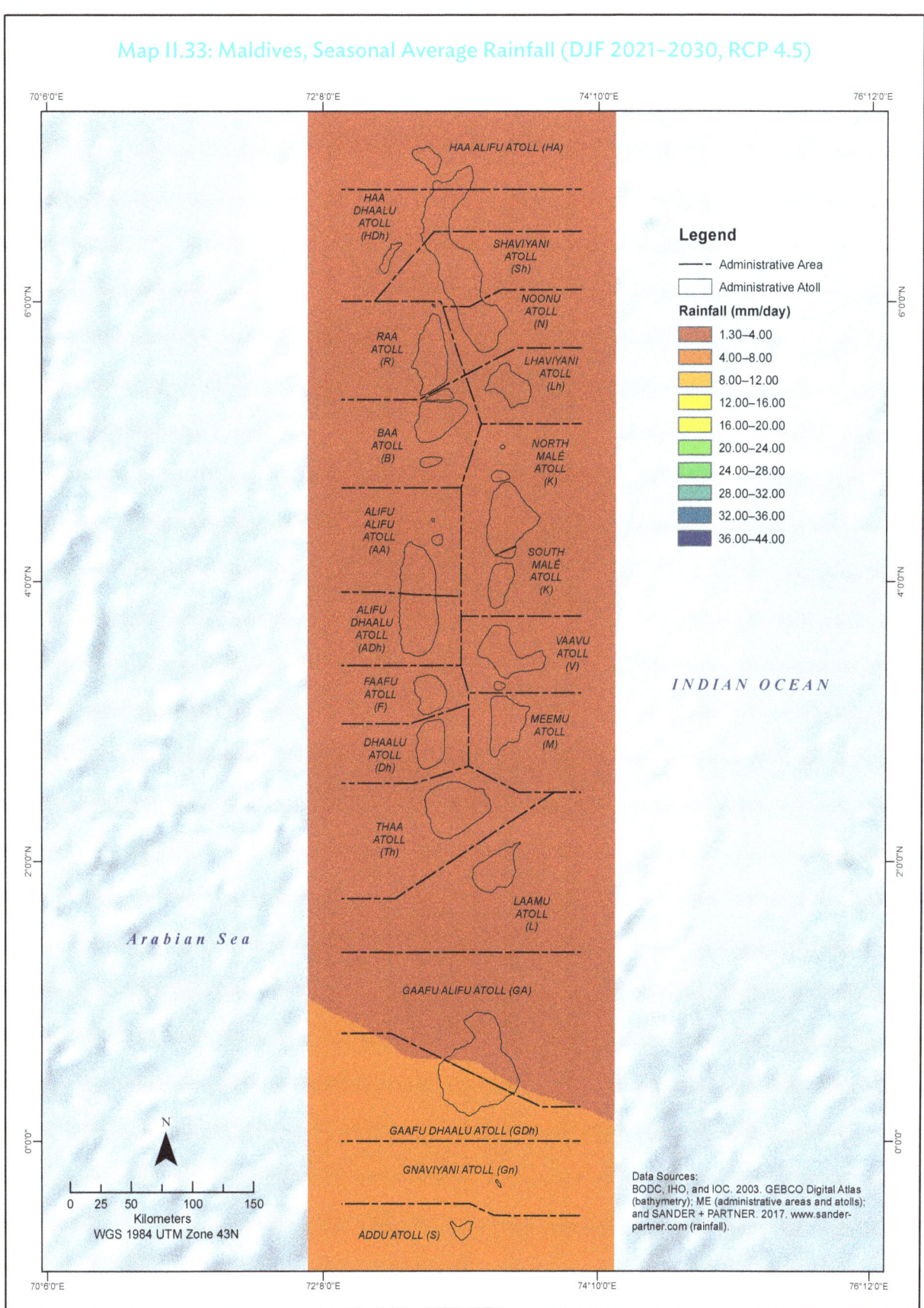
Map II.33: Maldives, Seasonal Average Rainfall (DJF 2021–2030, RCP 4.5)

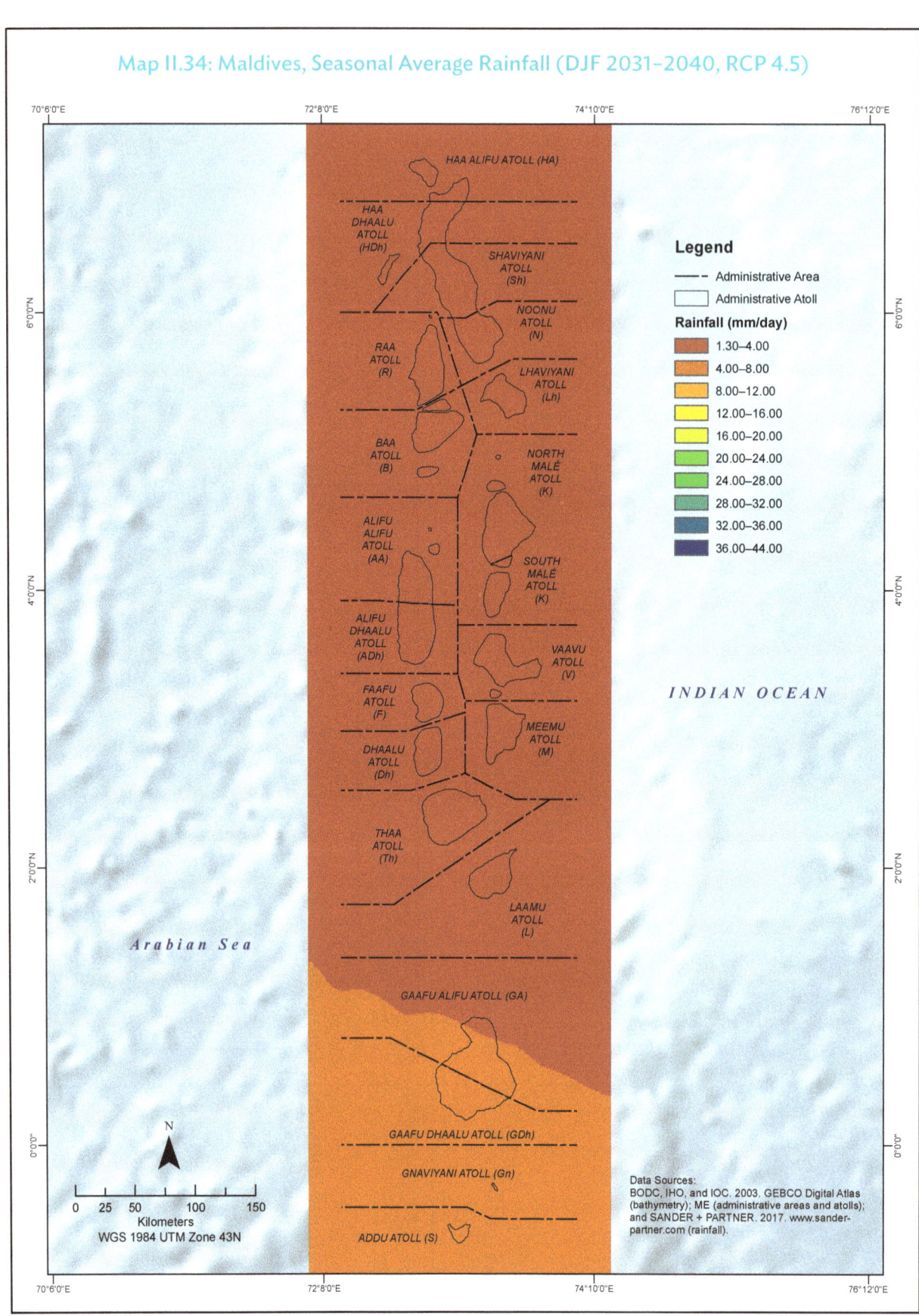

Future Climate 41

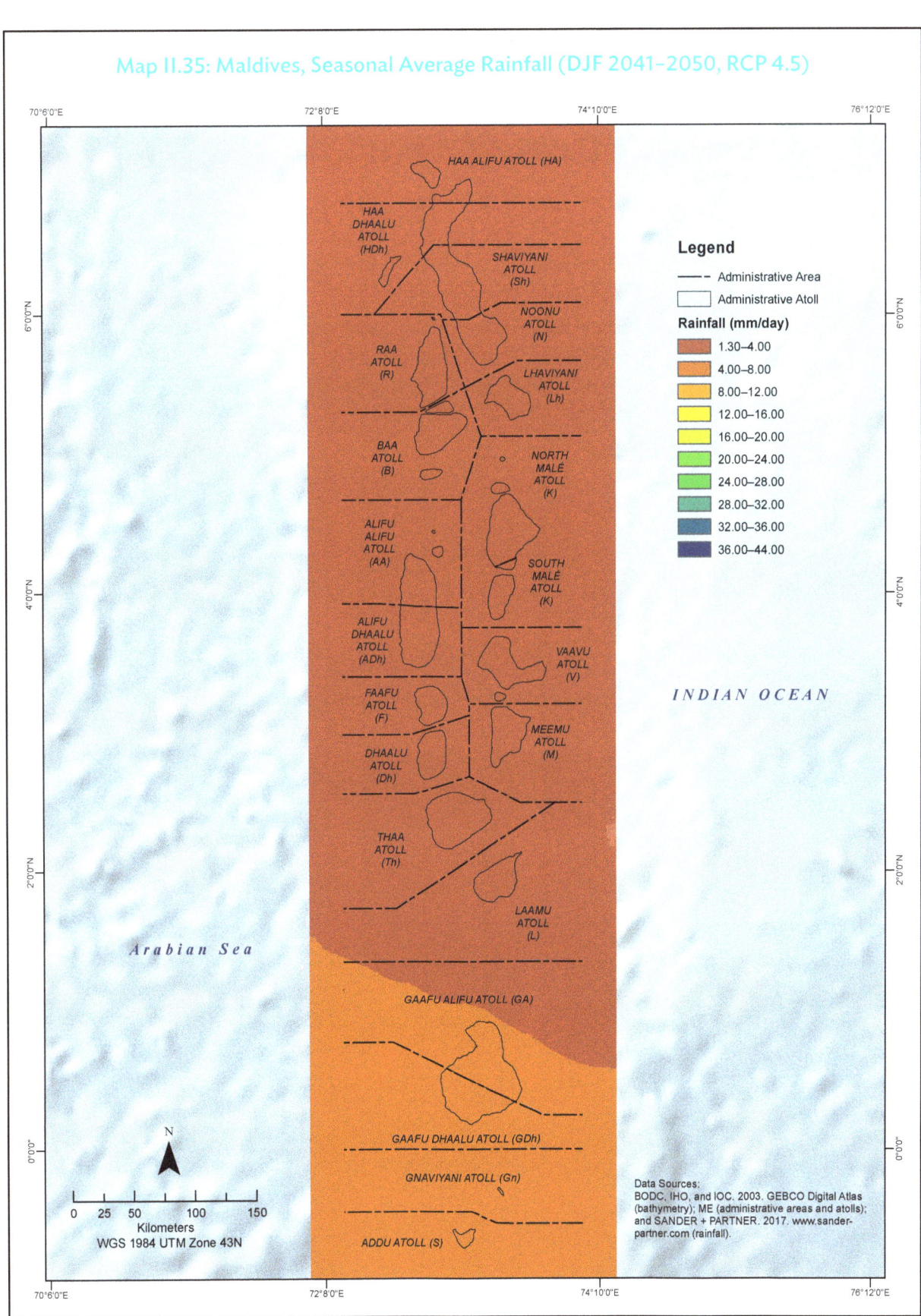

Map II.35: Maldives, Seasonal Average Rainfall (DJF 2041–2050, RCP 4.5)

Average Seasonal Rainfall Projection (MAM, RCP 4.5)

- ☑ Generally dry
- ☑ Driest season
- ☑ No significant change across the decades

Maldives experiences the driest days during MAM, as the northeast monsoon is strongest these months. Rainfall, at a rate of 1.3–4 mm/day, is distributed evenly throughout the atolls during MAM and throughout the projected time periods (with the exception of Haa Alifu Atoll in the 2030s).

Map II.36: Maldives, Average Seasonal Rainfall Projection (MAM, RCP 4.5)

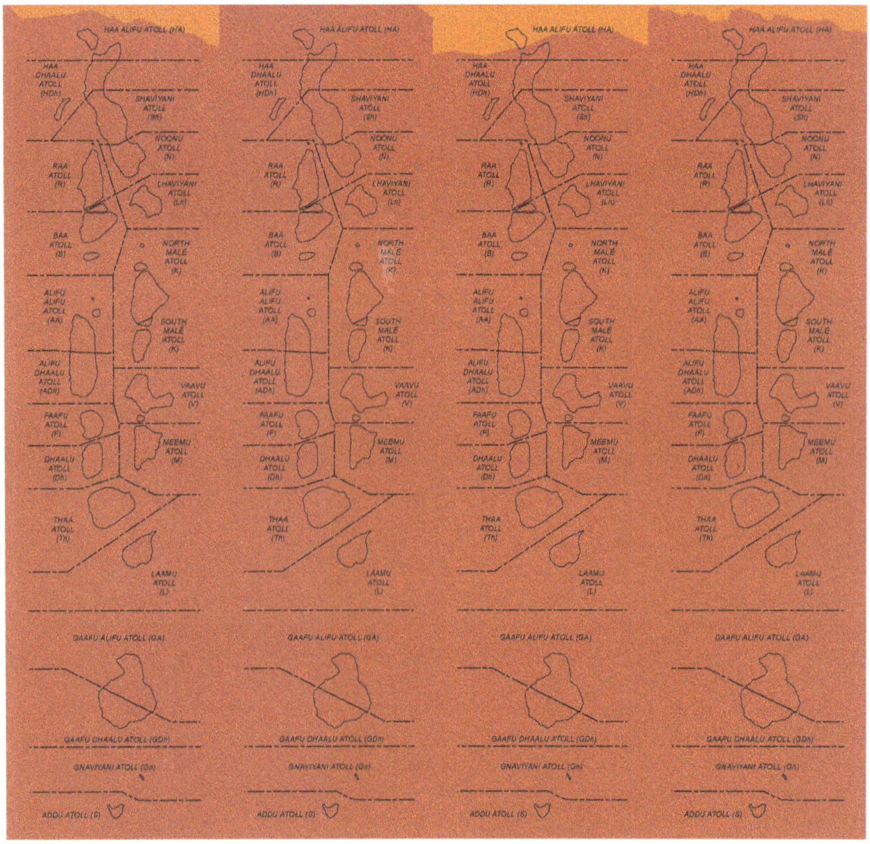

2011–2020 | 2021–2030 | 2031–2040 | 2041–2050

Legend
— Administrative Area
☐ Administrative Atoll

Rainfall (mm/day)
- 1.30–4.00
- 4.00–8.00
- 8.00–12.00
- 12.00–16.00
- 16.00–20.00
- 20.00–24.00
- 24.00–28.00
- 28.00–32.00
- 32.00–36.00
- 36.00–44.00

Data Sources:
BODC, IHO, and IOC. 2003.
GEBCO Digital Atlas (bathymetry);
ME (administrative areas and atoll);
and SANDER + PARTNER. 2017.
www.sander-partner.com (rainfall).

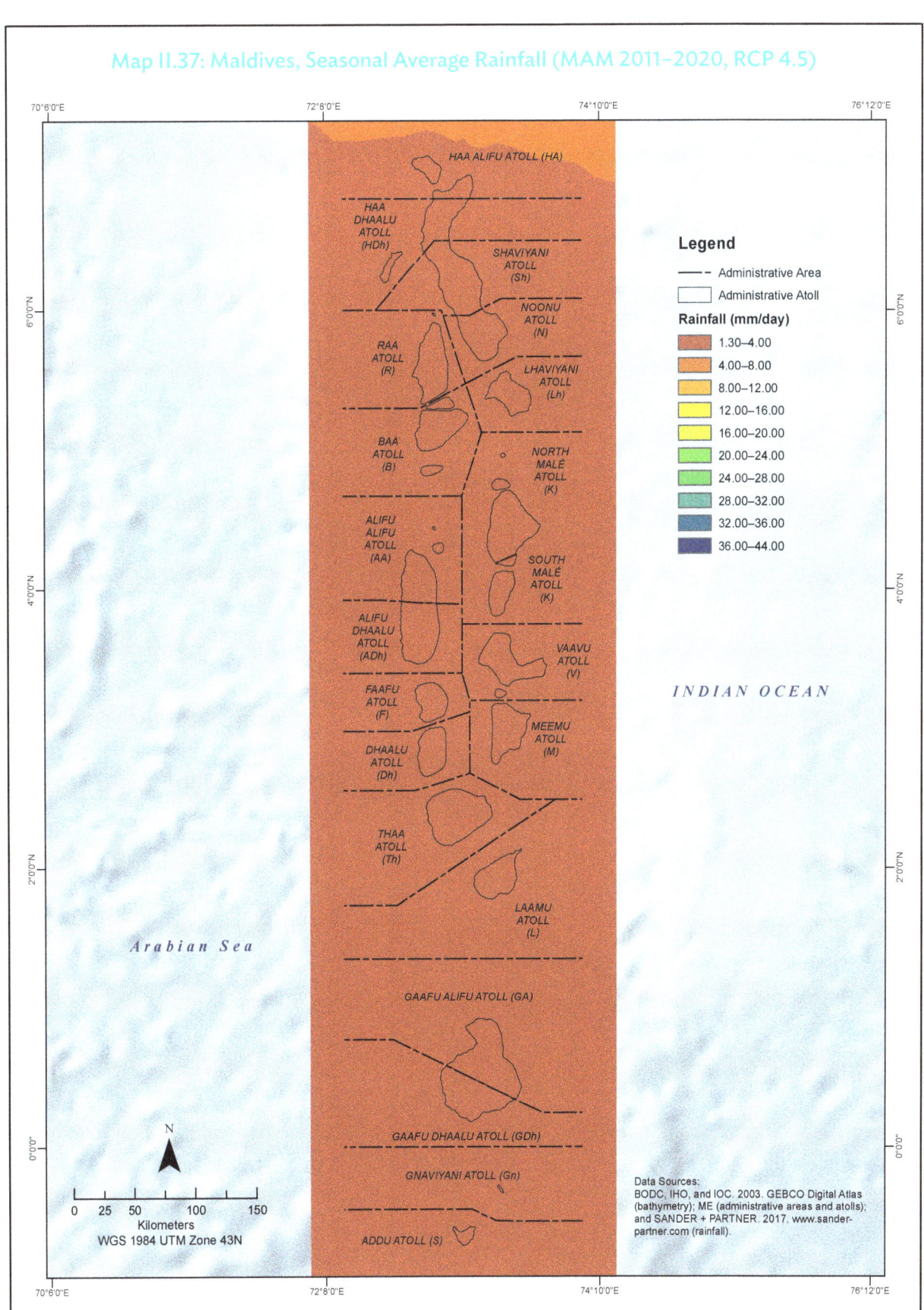

Map II.37: Maldives, Seasonal Average Rainfall (MAM 2011–2020, RCP 4.5)

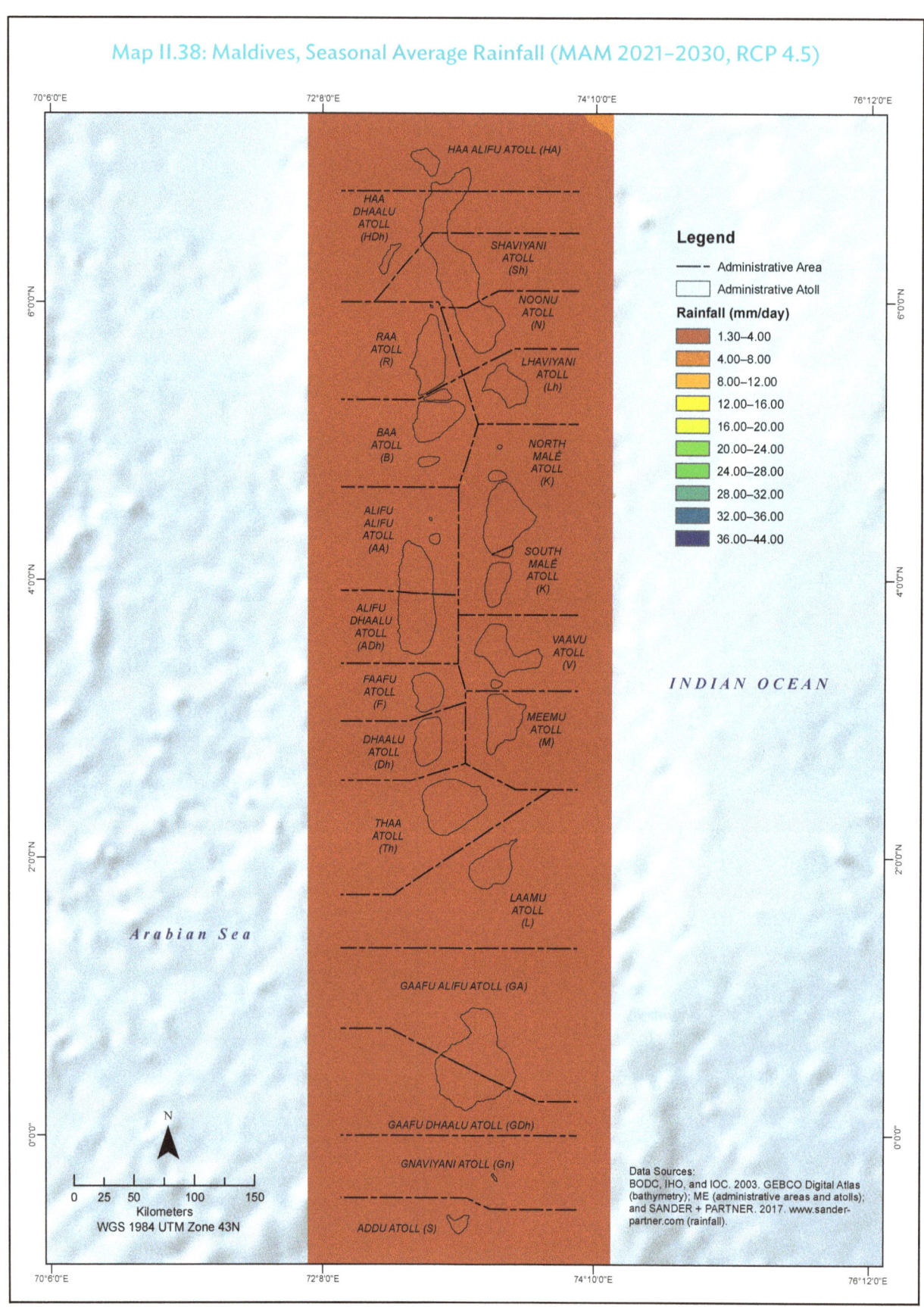

Future Climate

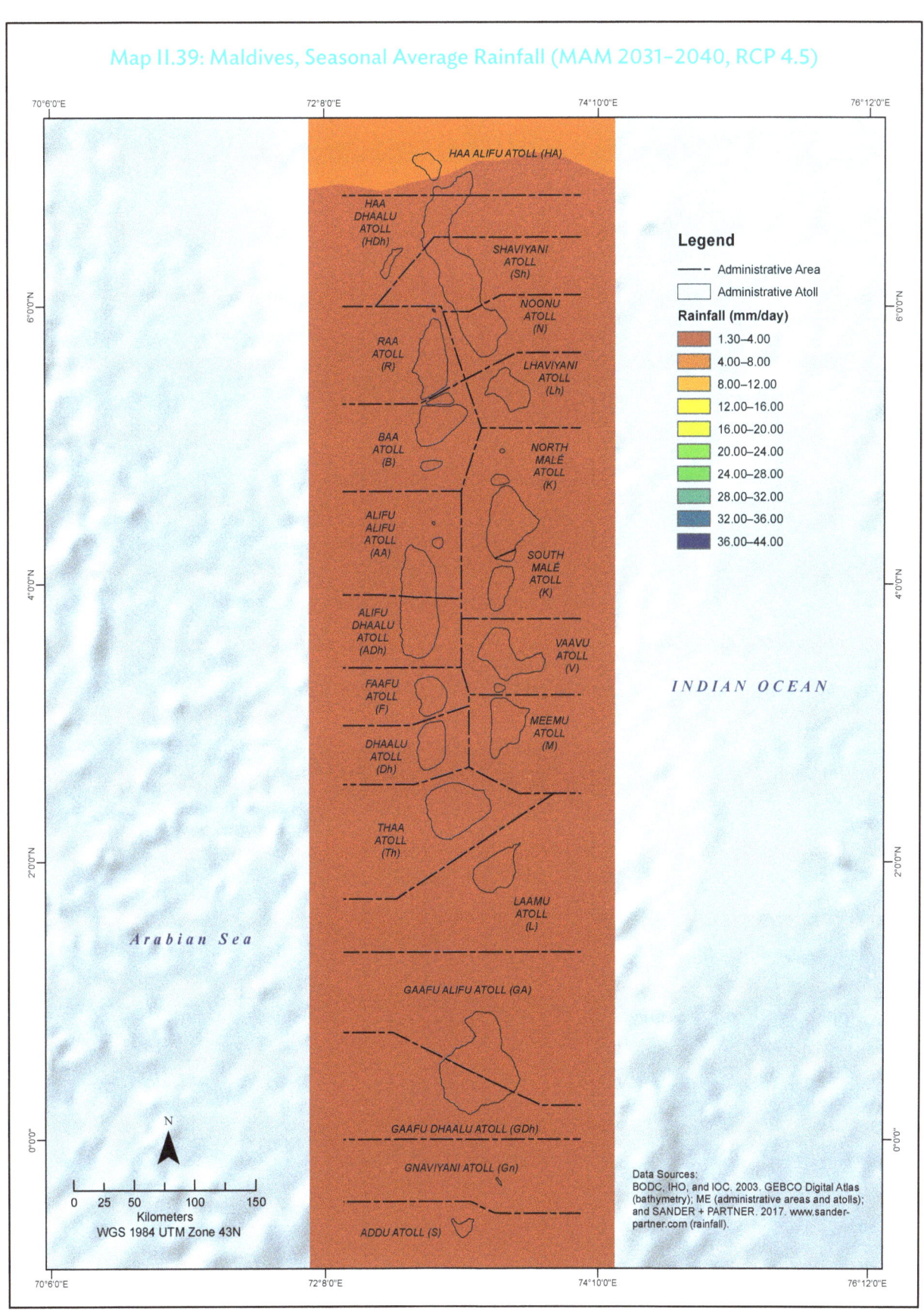

Map II.39: Maldives, Seasonal Average Rainfall (MAM 2031–2040, RCP 4.5)

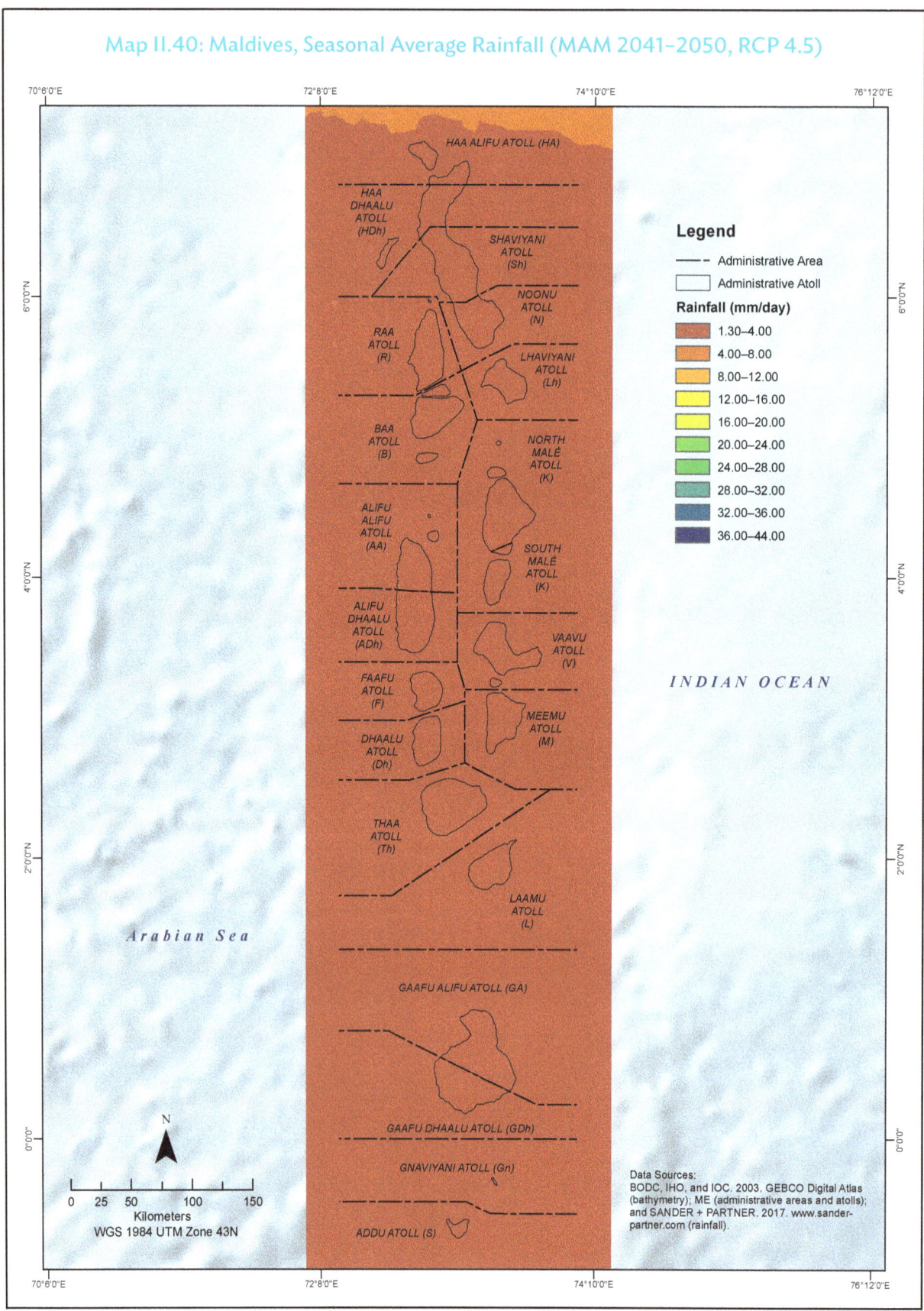

Average Seasonal Rainfall Projection (JJA, RCP 4.5)

47

Maldives experiences the highest rainfall rate during JJA compared with other seasons. In these months, rainfall is observed to increase, especially in the northern atolls as the southwest monsoon intensifies. Rainfall rate greater than 4 mm/day is expected to move southward in the 2030s, reaching the atolls in the middle portion of Maldives.

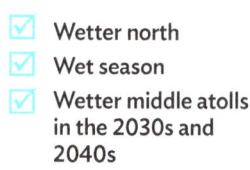

- ✓ Wetter north
- ✓ Wet season
- ✓ Wetter middle atolls in the 2030s and 2040s

Map II.41: Maldives, Average Seasonal Rainfall Projection (JJA, RCP 4.5)

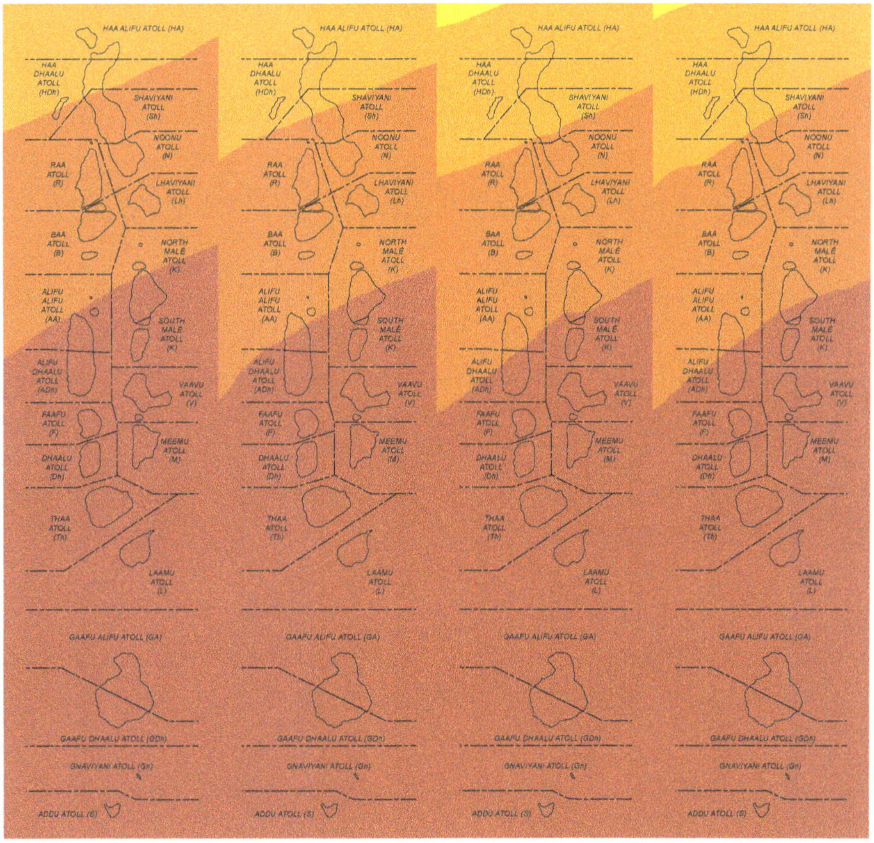

2011–2020　2021–2030　2031–2040　2041–2050

Legend

— — Administrative Area

☐ Administrative Atoll

Rainfall (mm/day)

- 1.30–4.00
- 4.00–8.00
- 8.00–12.00
- 12.00–16.00
- 16.00–20.00
- 20.00–24.00
- 24.00–28.00
- 28.00–32.00
- 32.00–36.00
- 36.00–44.00

Data Sources:
BODC, IHO, and IOC. 2003.
GEBCO Digital Atlas (bathymetry);
ME (administrative areas and atoll);
and SANDER + PARTNER. 2017.
www.sander-partner.com (rainfall).

48 Multihazard Risk Atlas of Maldives—Climate and Geophysical Hazards

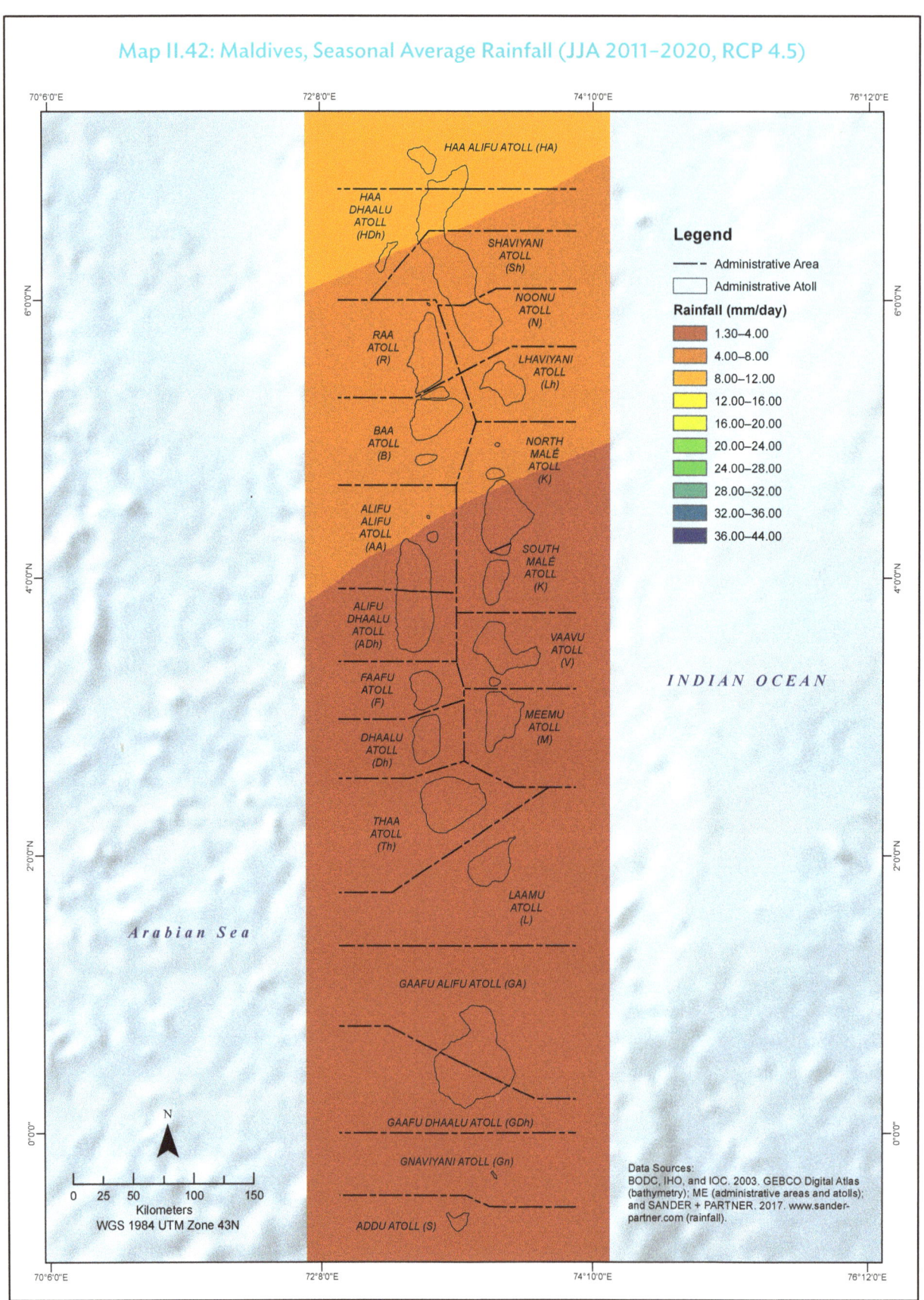

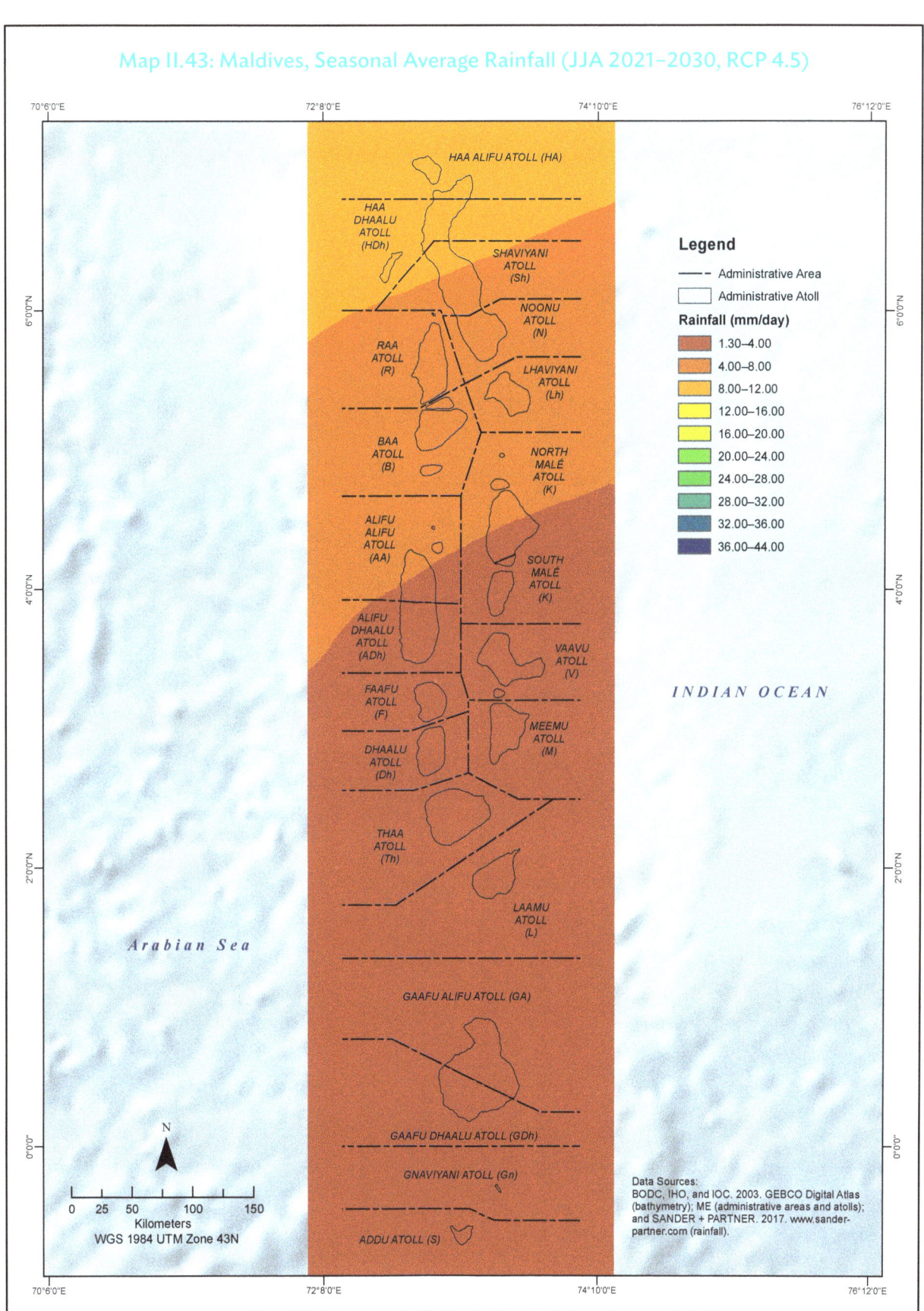

50 Multihazard Risk Atlas of Maldives—Climate and Geophysical Hazards

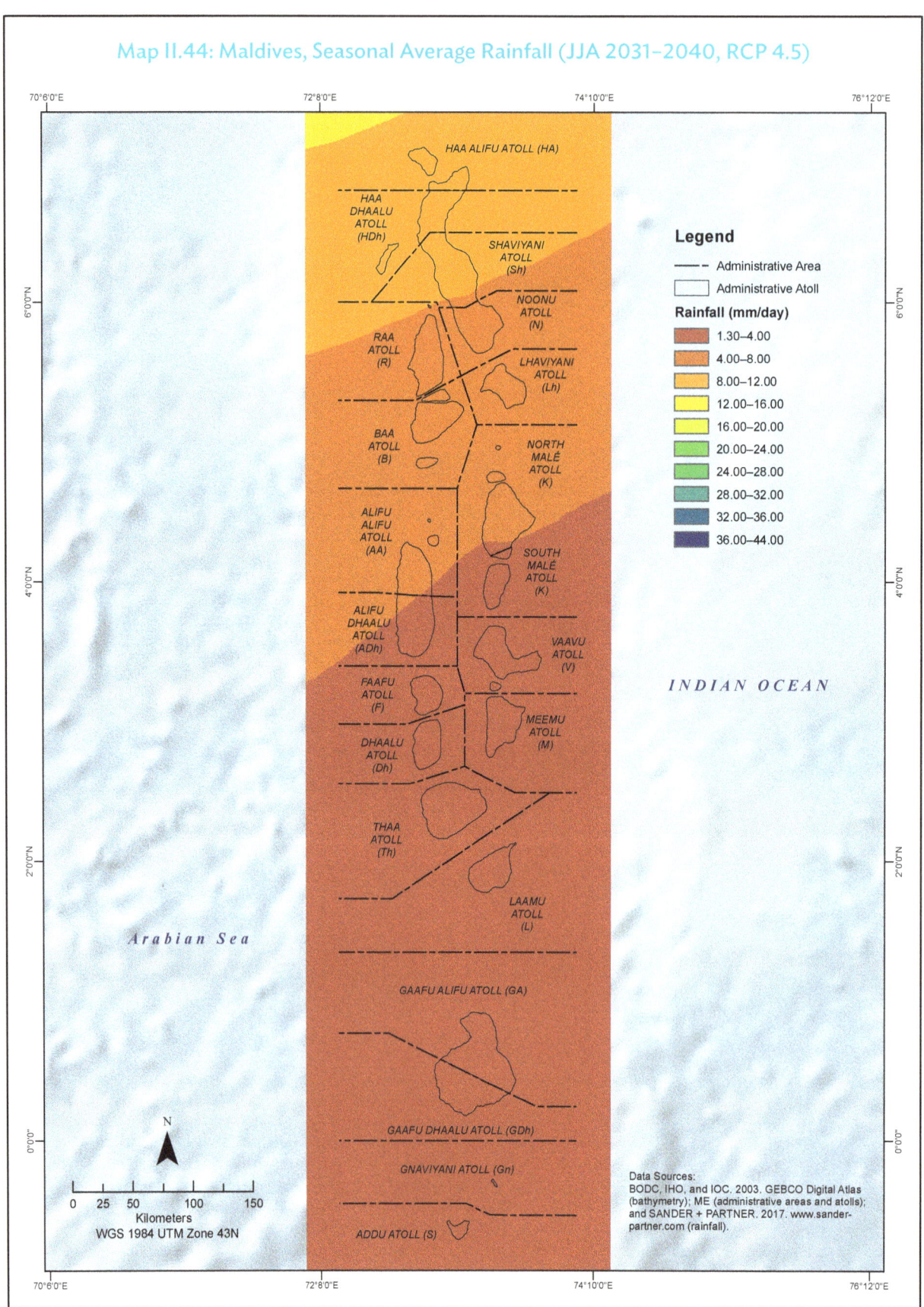

Future Climate 51

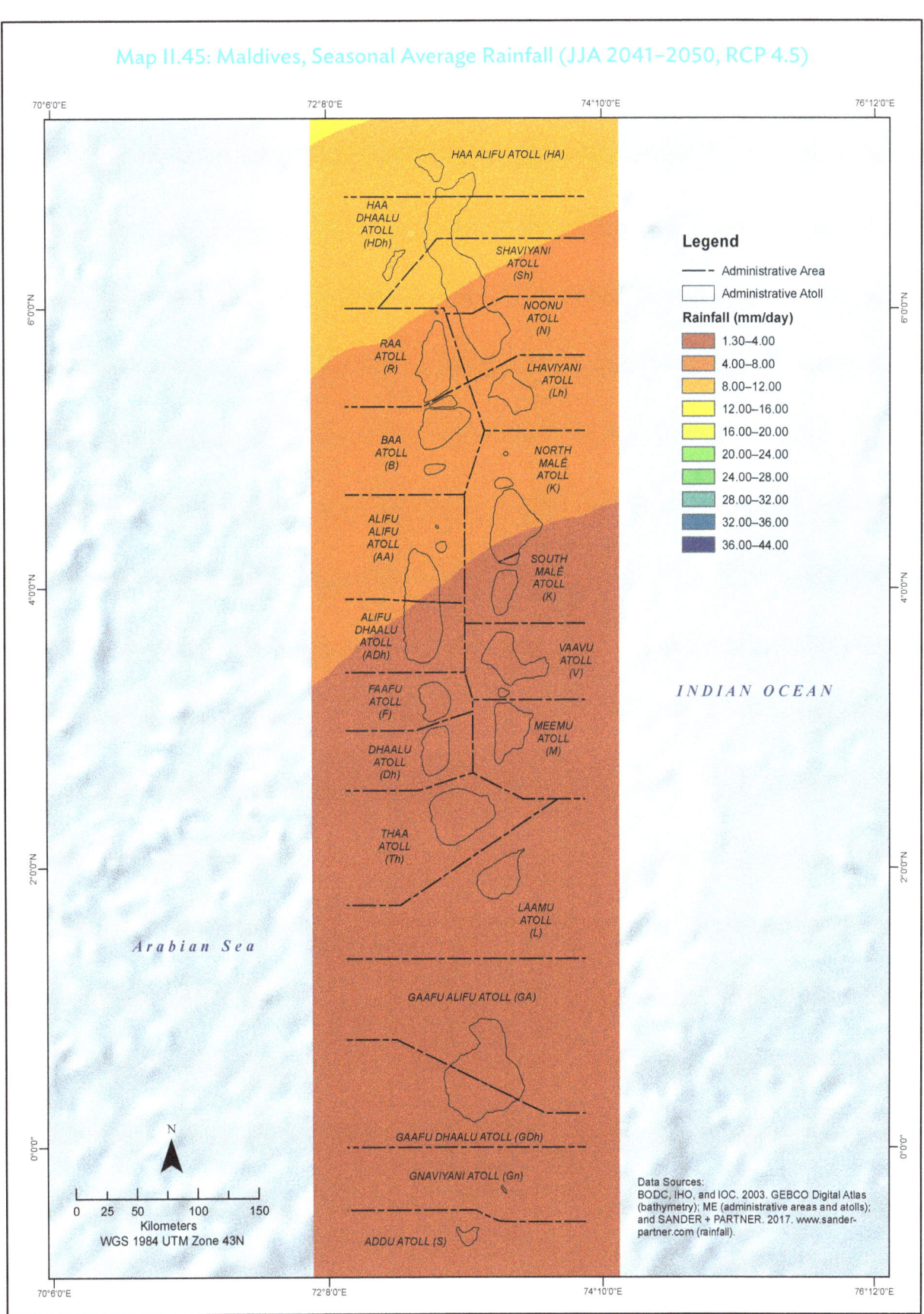

Average Seasonal Rainfall Projection (SON, RCP 4.5)

- ☑ Wetter north
- ☑ Wet season
- ☑ No significant shift in rainfall distribution across decades

During SON, the southwest monsoon starts to weaken, bringing the rainfall rate down to less than 8 mm/day. The maps presenting the time series of seasonal rainfall for SON consistently show that across the projected time periods, the southern half of Maldives receives minimal (1.3–4 mm/day) rainfall while the northern half of Maldives experiences a higher rainfall rate (4–8 mm/day).

Map II.46: Maldives, Average Seasonal Rainfall Projection (SON, RCP 4.5)

Legend
- Administrative Area
- Administrative Atoll

Rainfall (mm/day)
- 1.30–4.00
- 4.00–8.00
- 8.00–12.00
- 12.00–16.00
- 16.00–20.00
- 20.00–24.00
- 24.00–28.00
- 28.00–32.00
- 32.00–36.00
- 36.00–44.00

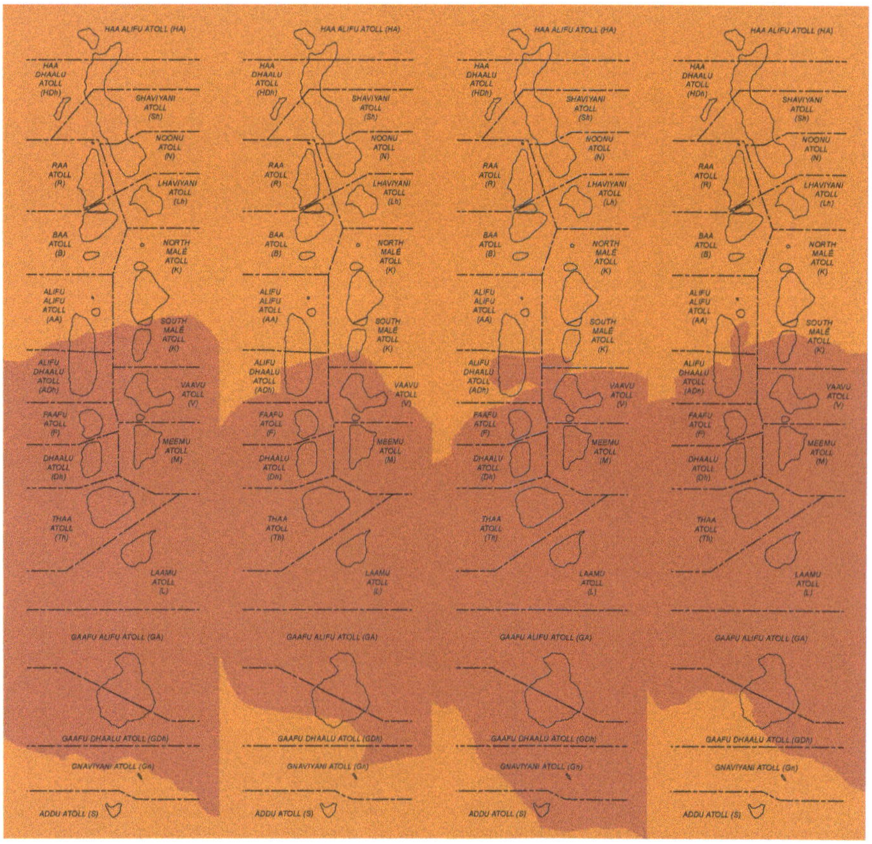

Data Sources: BODC, IHO, and IOC. 2003. GEBCO Digital Atlas (bathymetry); ME (administrative areas and atoll); and SANDER + PARTNER. 2017. www.sander-partner.com (rainfall).

2011–2020 | 2021–2030 | 2031–2040 | 2041–2050

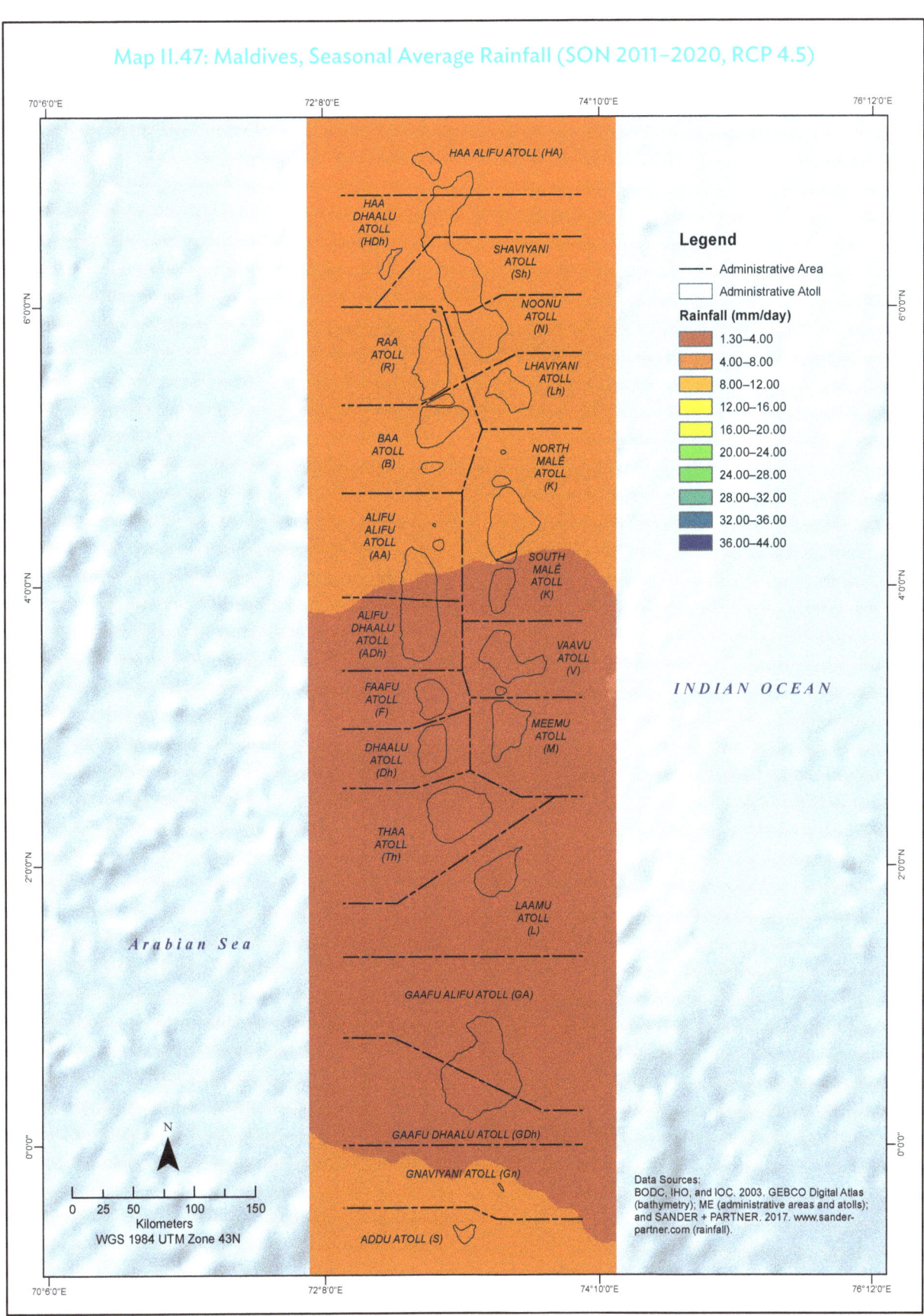

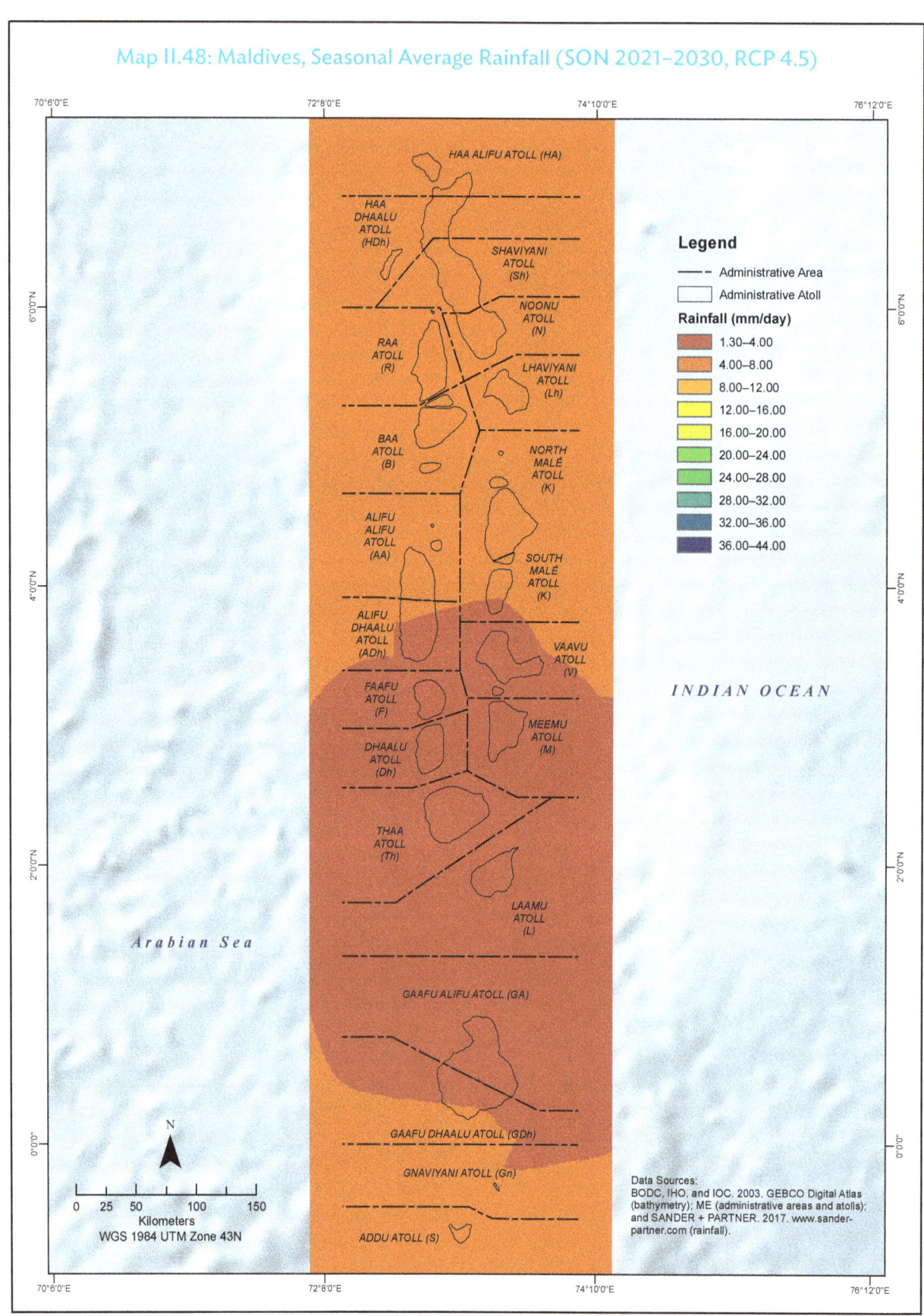

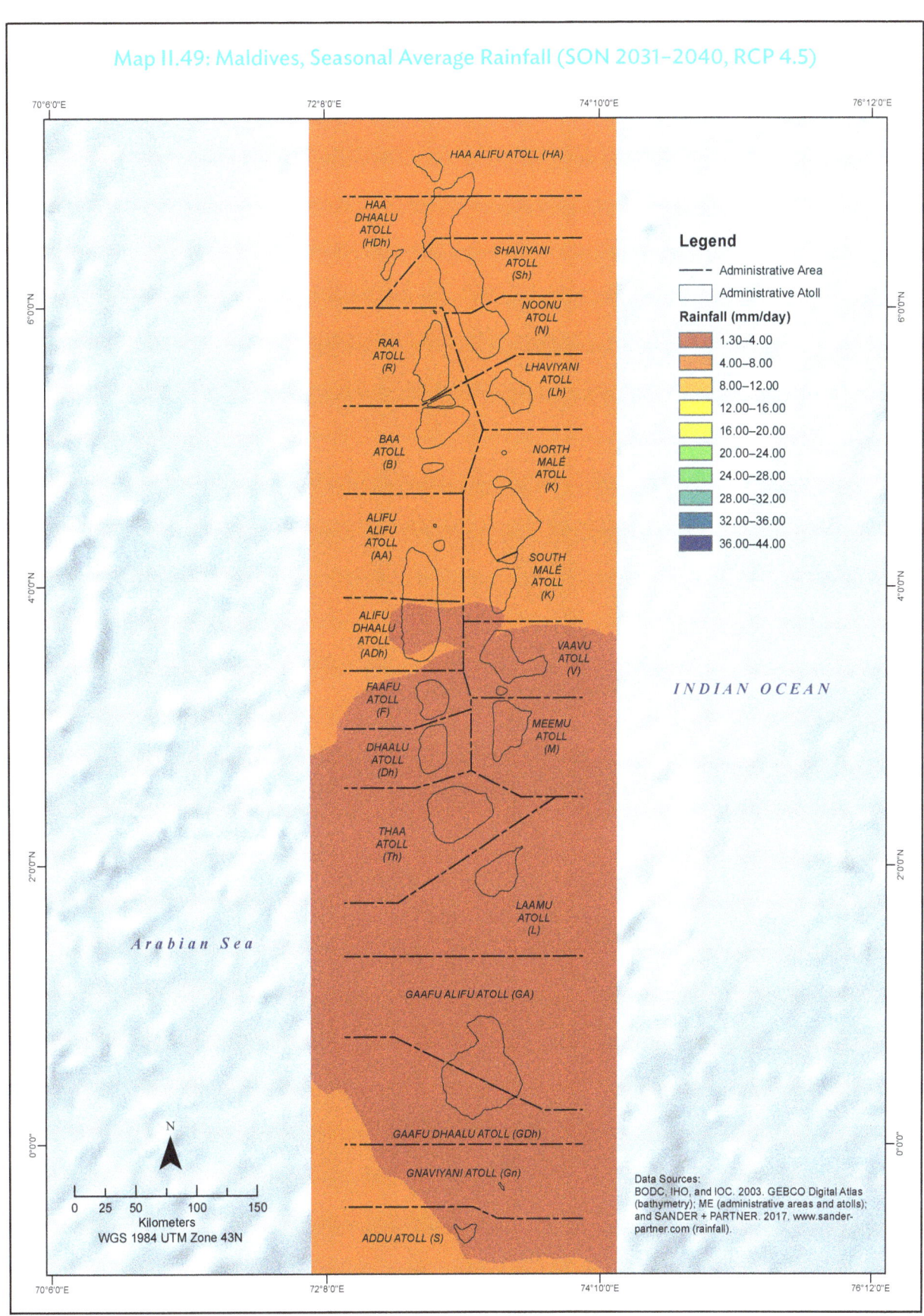

56 Multihazard Risk Atlas of Maldives—Climate and Geophysical Hazards

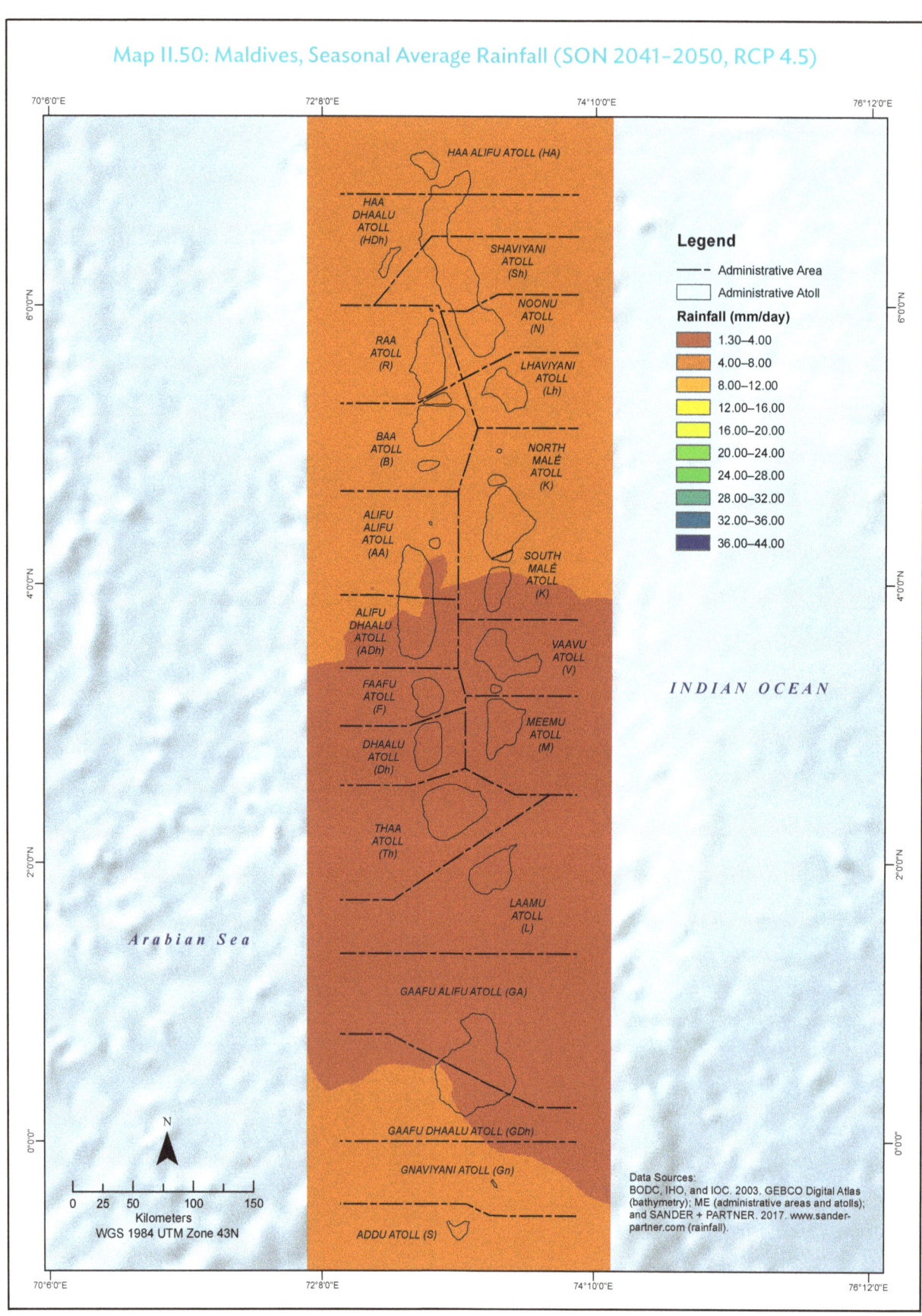

Map II.50: Maldives, Seasonal Average Rainfall (SON 2041–2050, RCP 4.5)

Average Annual Rainfall Projection (RCP 8.5)

57

RCP 8.5 depicts a scenario in which greenhouse gas emissions are higher. As a result, the change in rainfall would also be greater compared with RCP 4.5. However, when comparing the RCP 8.5 and RCP 4.5 maps, there is minimal visible difference in the average rainfall. In this time series of annual rainfall projection for RCP 8.5, an increasing rainfall rate trend from 2011 to 2050 is visible. Higher rainfall rate (4–16 mm/day) will be experienced by atolls in northern Maldives. However, as the years progress, more atolls in the north will experience wetter days.

- ✓ Wetter north
- ✓ North will be wetter in the future
- ✓ No significant difference with RCP 4.5

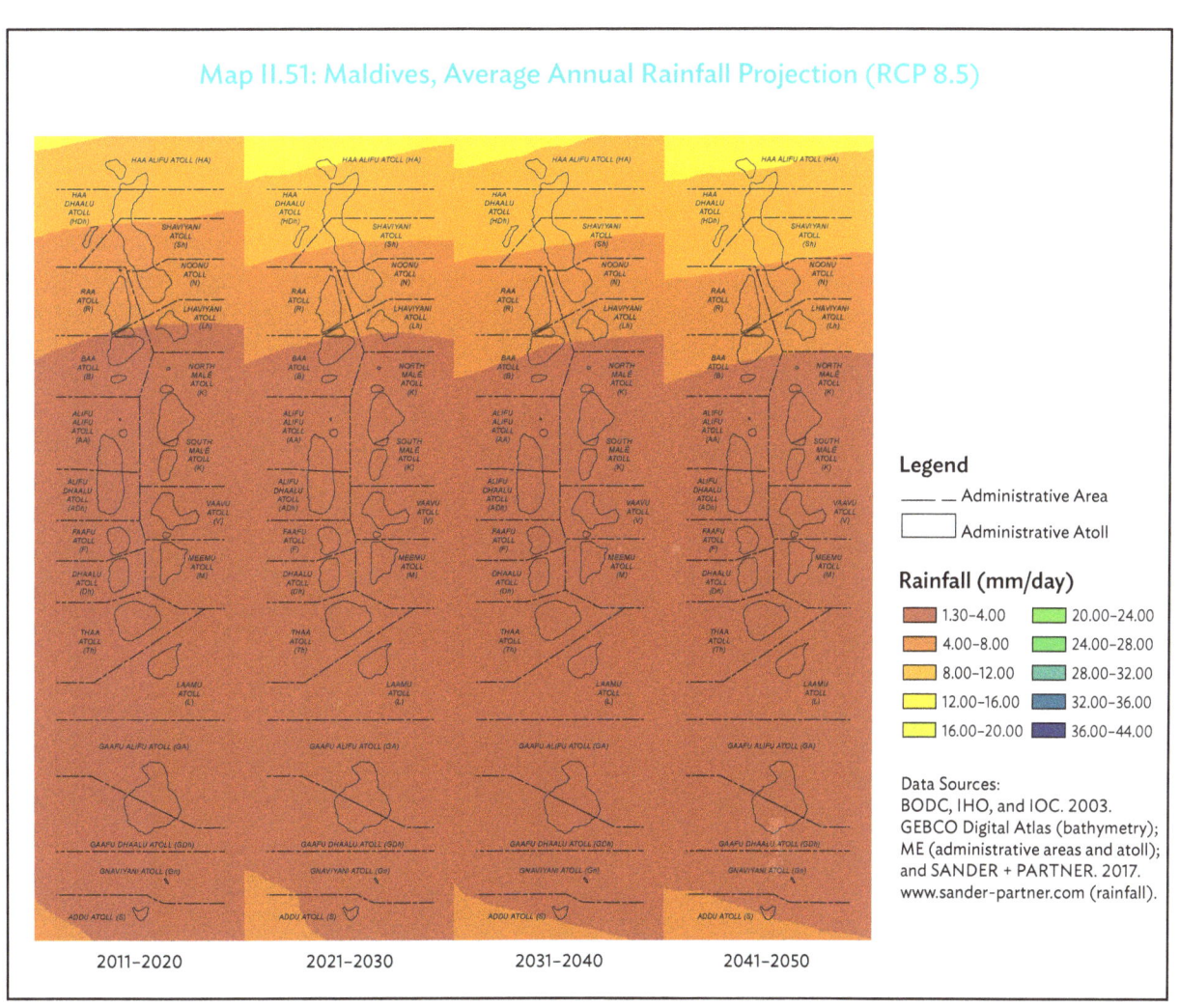

Map II.51: Maldives, Average Annual Rainfall Projection (RCP 8.5)

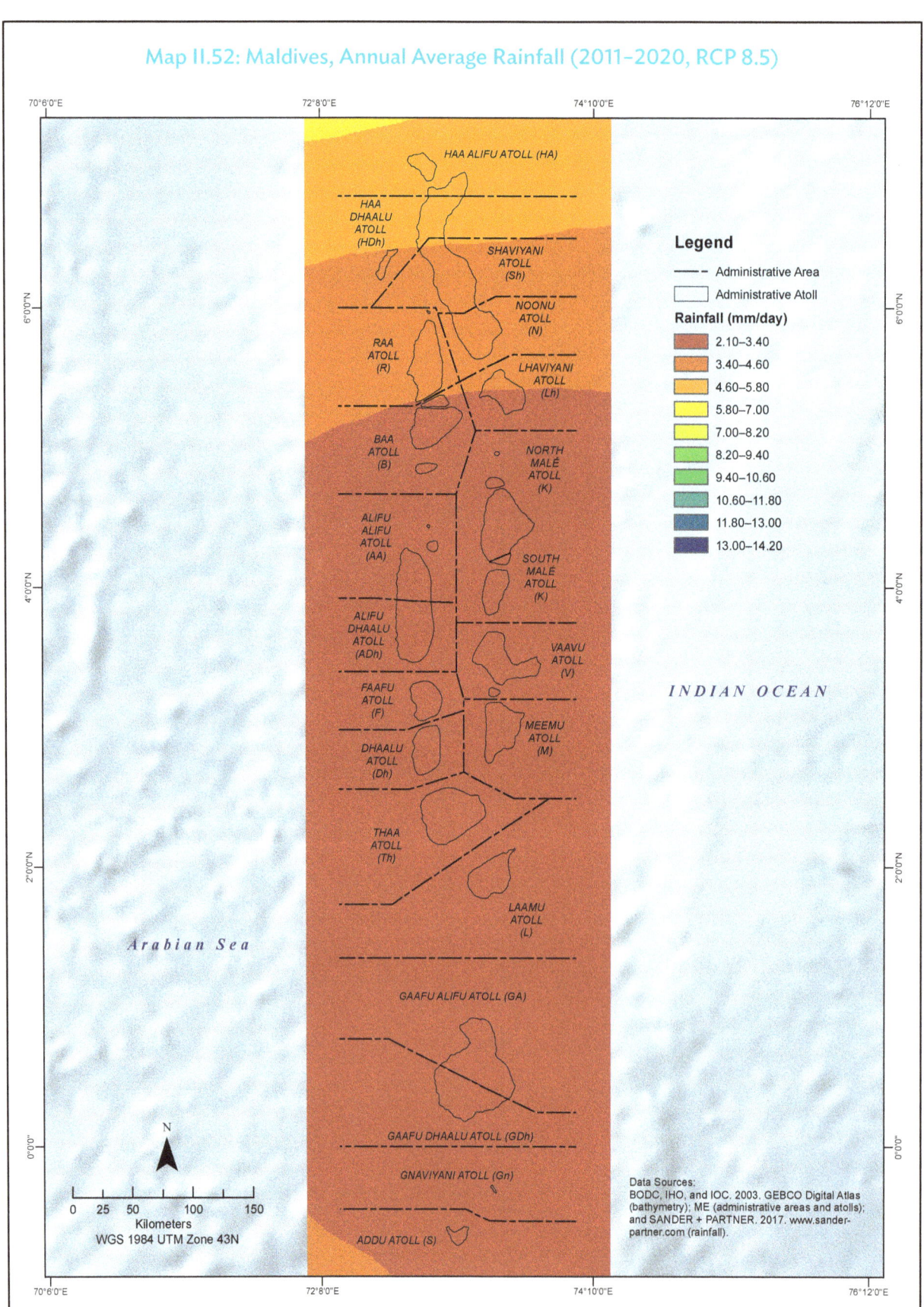

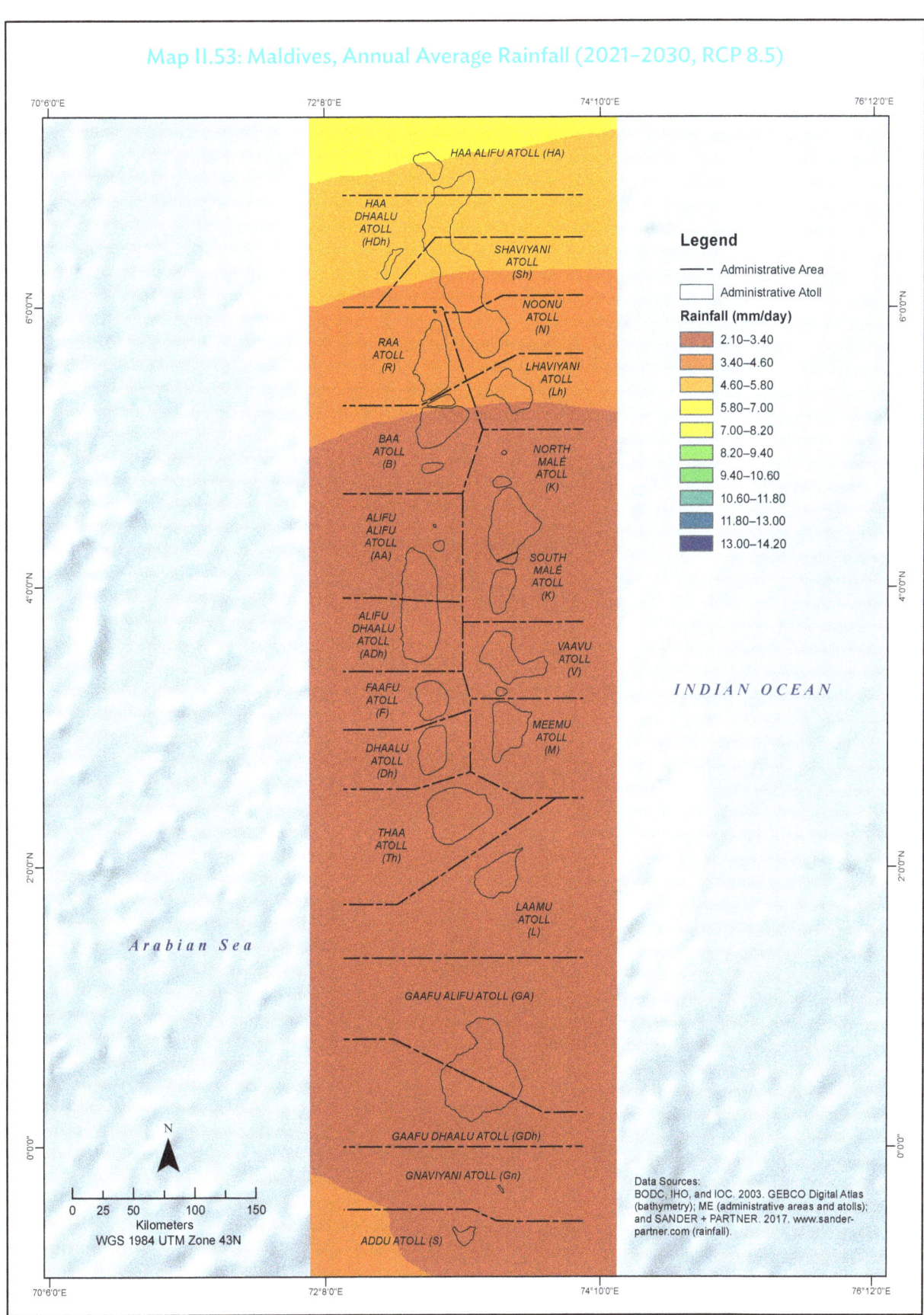

Map II.53: Maldives, Annual Average Rainfall (2021–2030, RCP 8.5)

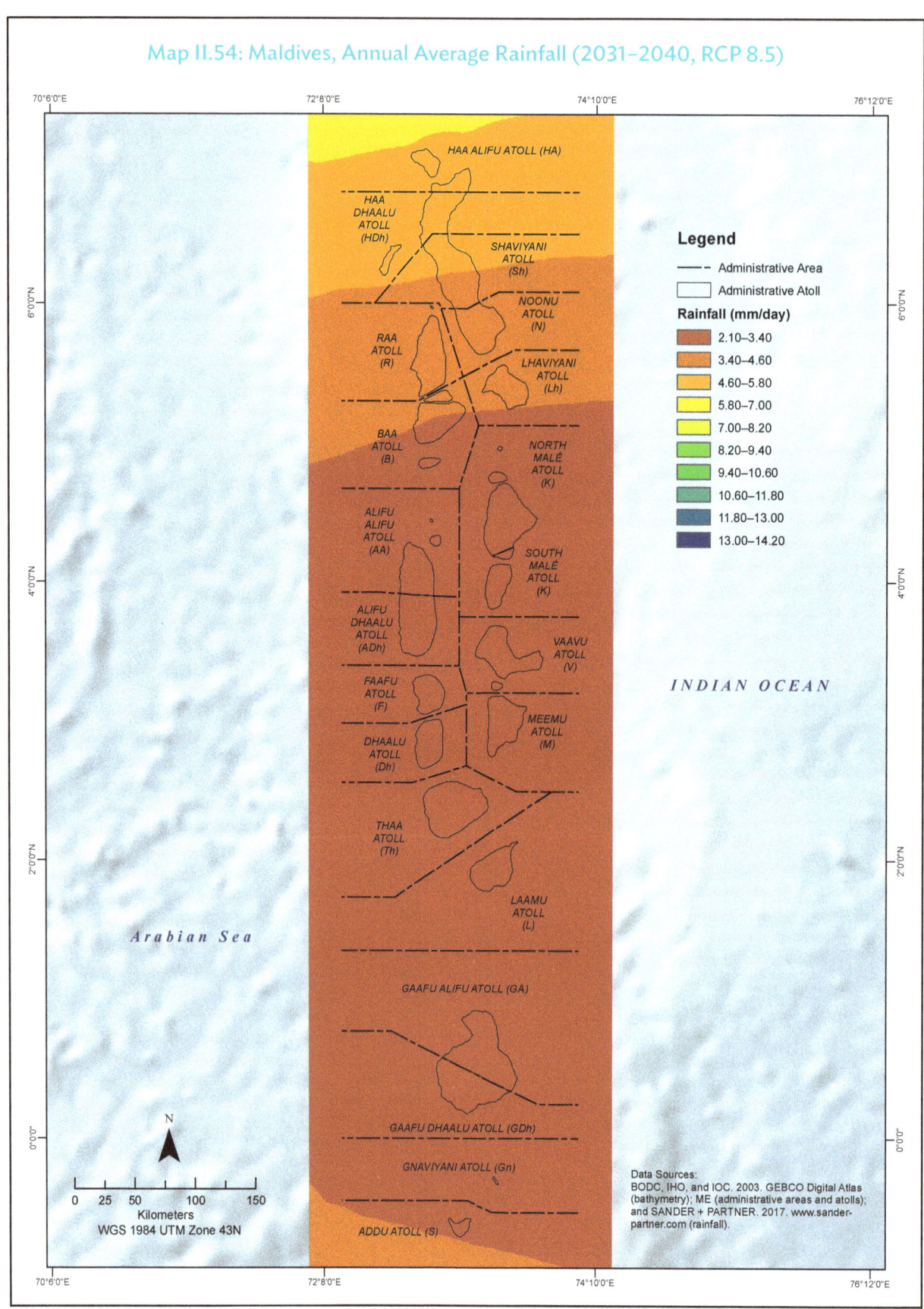

Map II.54: Maldives, Annual Average Rainfall (2031–2040, RCP 8.5)

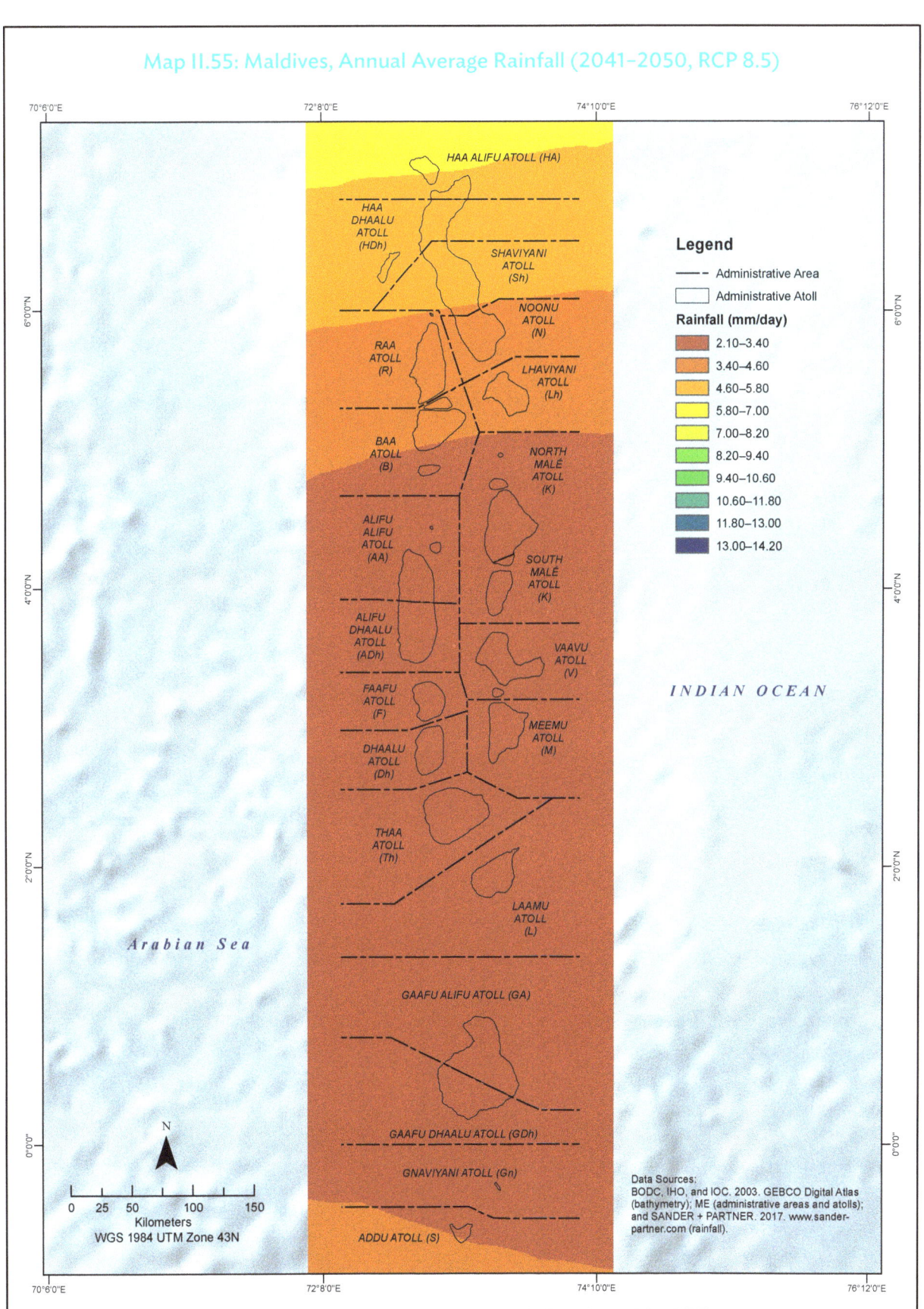

Map II.55: Maldives, Annual Average Rainfall (2041–2050, RCP 8.5)

Average Seasonal Rainfall Projection (DJF, RCP 8.5)

- ✓ Wetter south
- ✓ More atolls in the south will experience higher rainfall rate
- ✓ 4–8 mm/day rainfall rate has greater coverage than RCP 4.5 projection

RCP 8.5 and RCP 4.5 projections of average seasonal rainfall show the same picture with a slight difference. The south of Maldives will remain wetter compared to the north. More atolls in the south will experience a higher rainfall rate (4–8 mm/day). However, there is a slight increase in the area covered by the 4–8 mm/day rainfall rate for RCP 8.5 as compared to the RCP 4.5 projection.

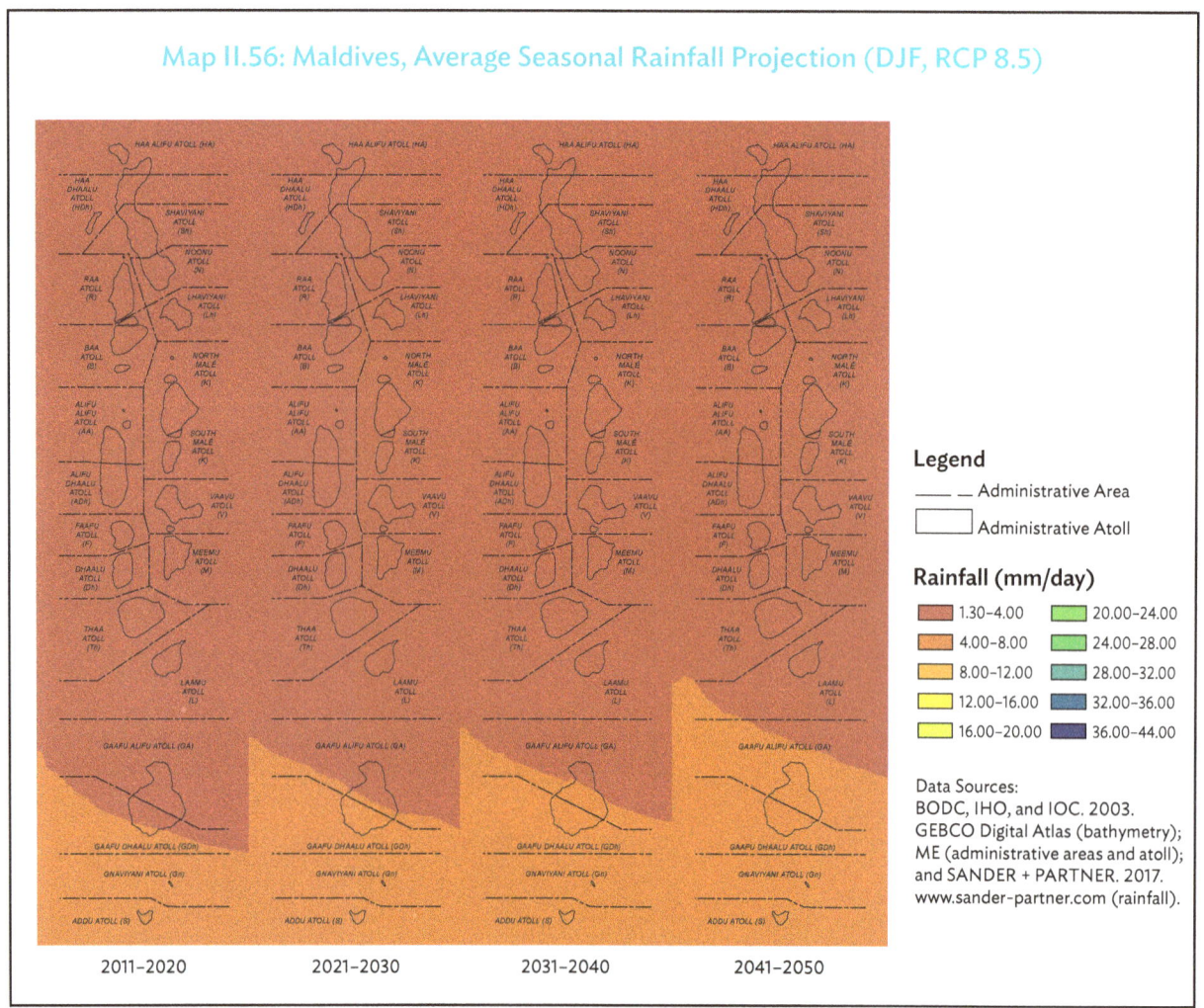

Map II.56: Maldives, Average Seasonal Rainfall Projection (DJF, RCP 8.5)

Future Climate 63

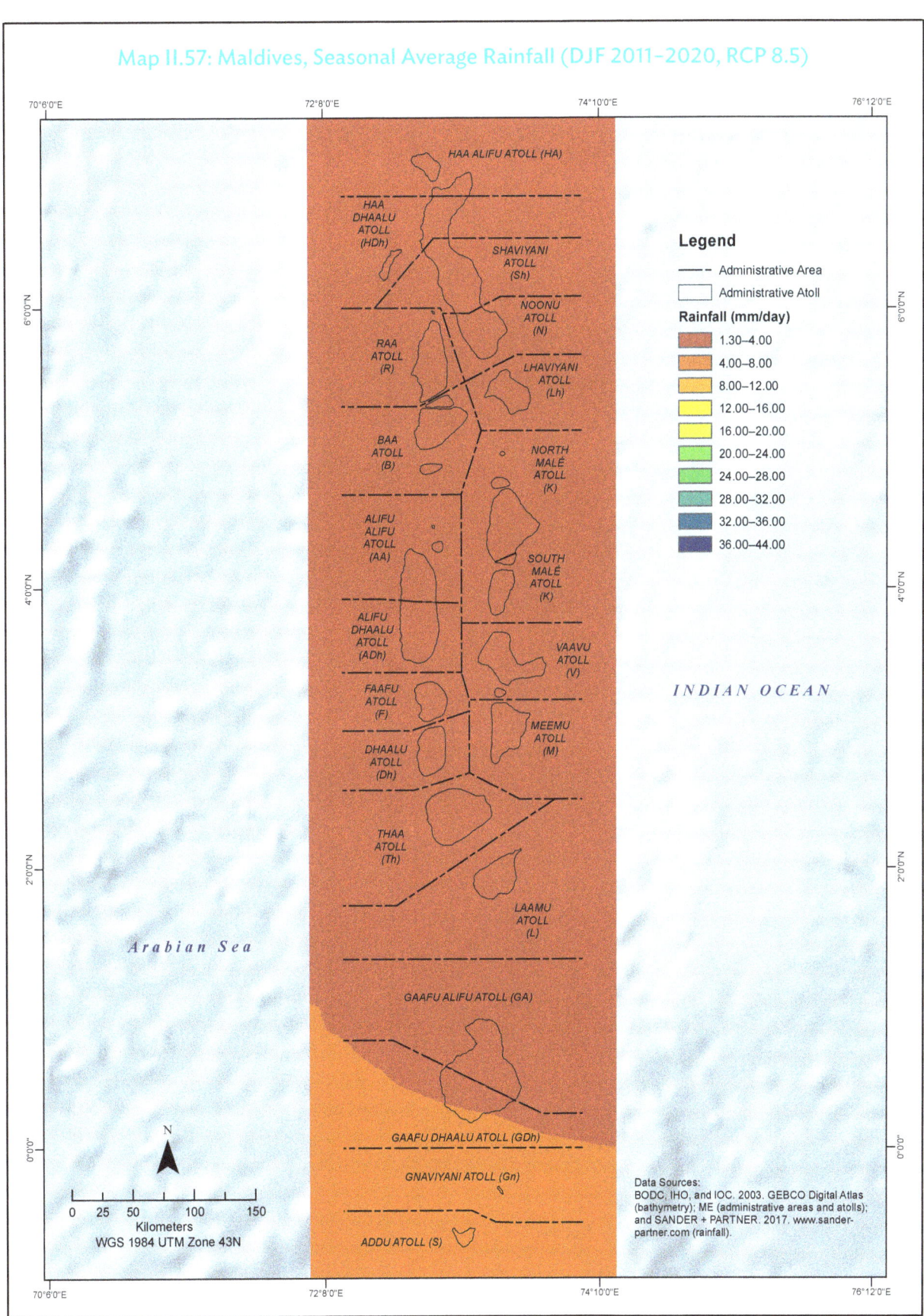

64 Multihazard Risk Atlas of Maldives—Climate and Geophysical Hazards

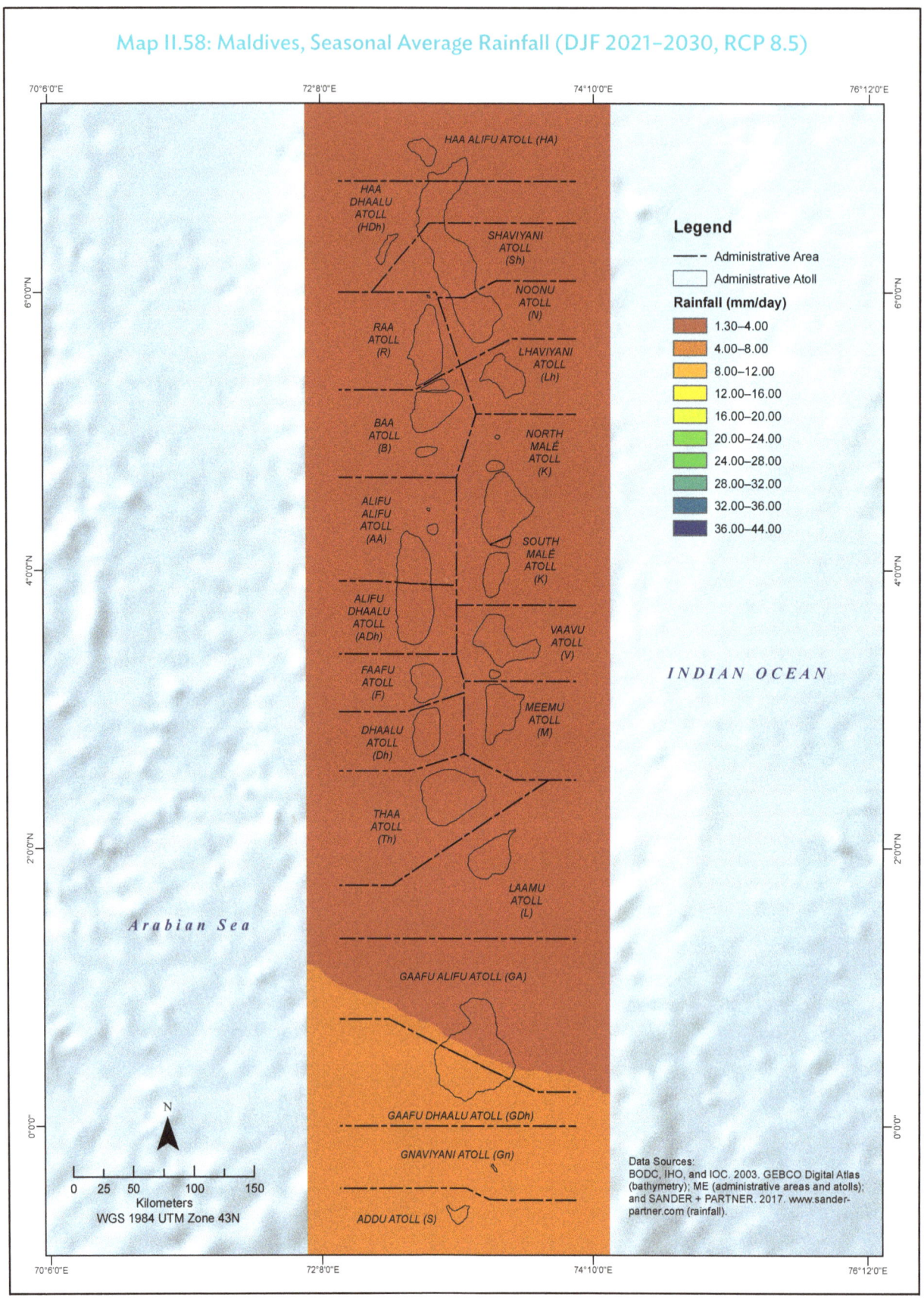

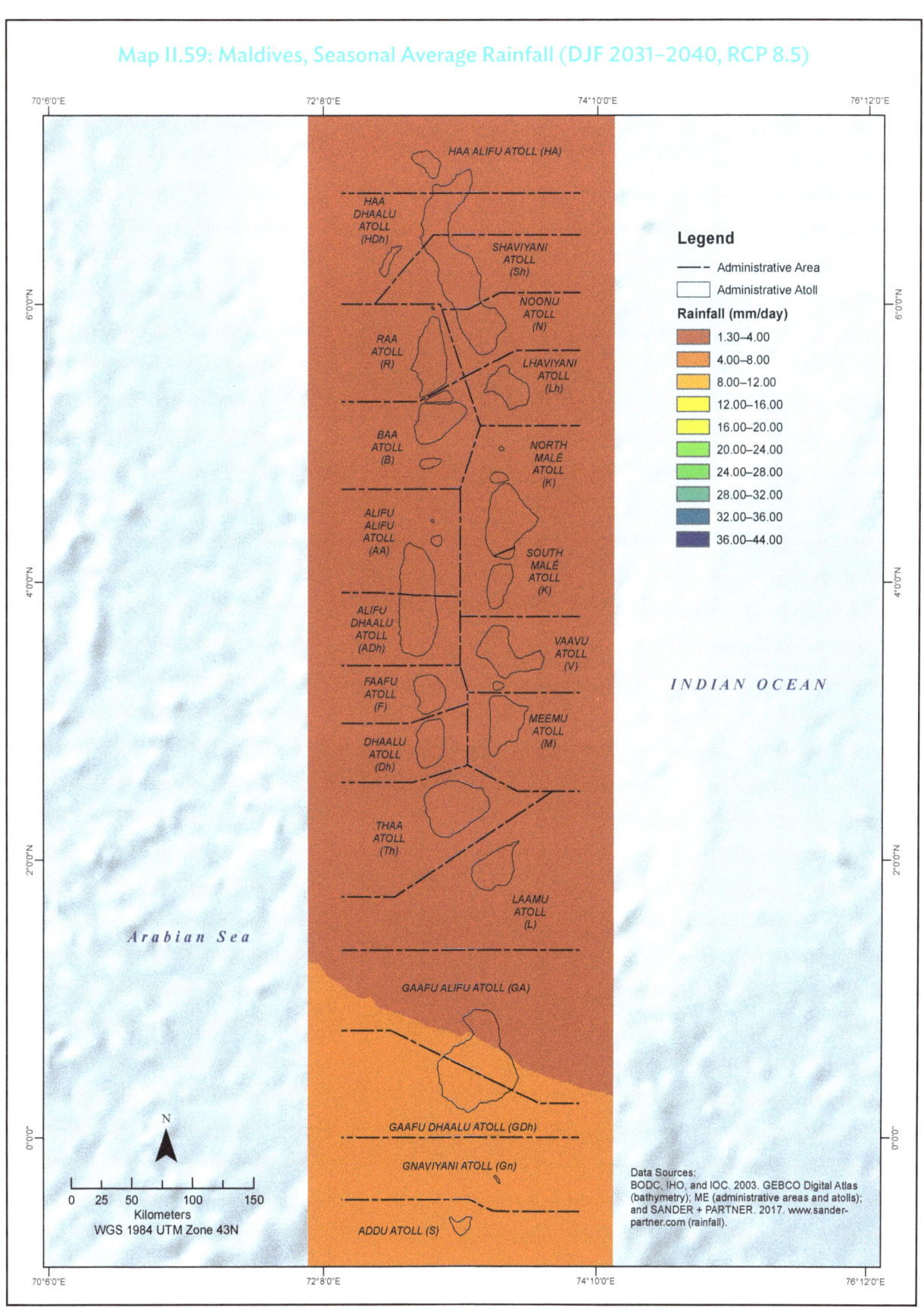

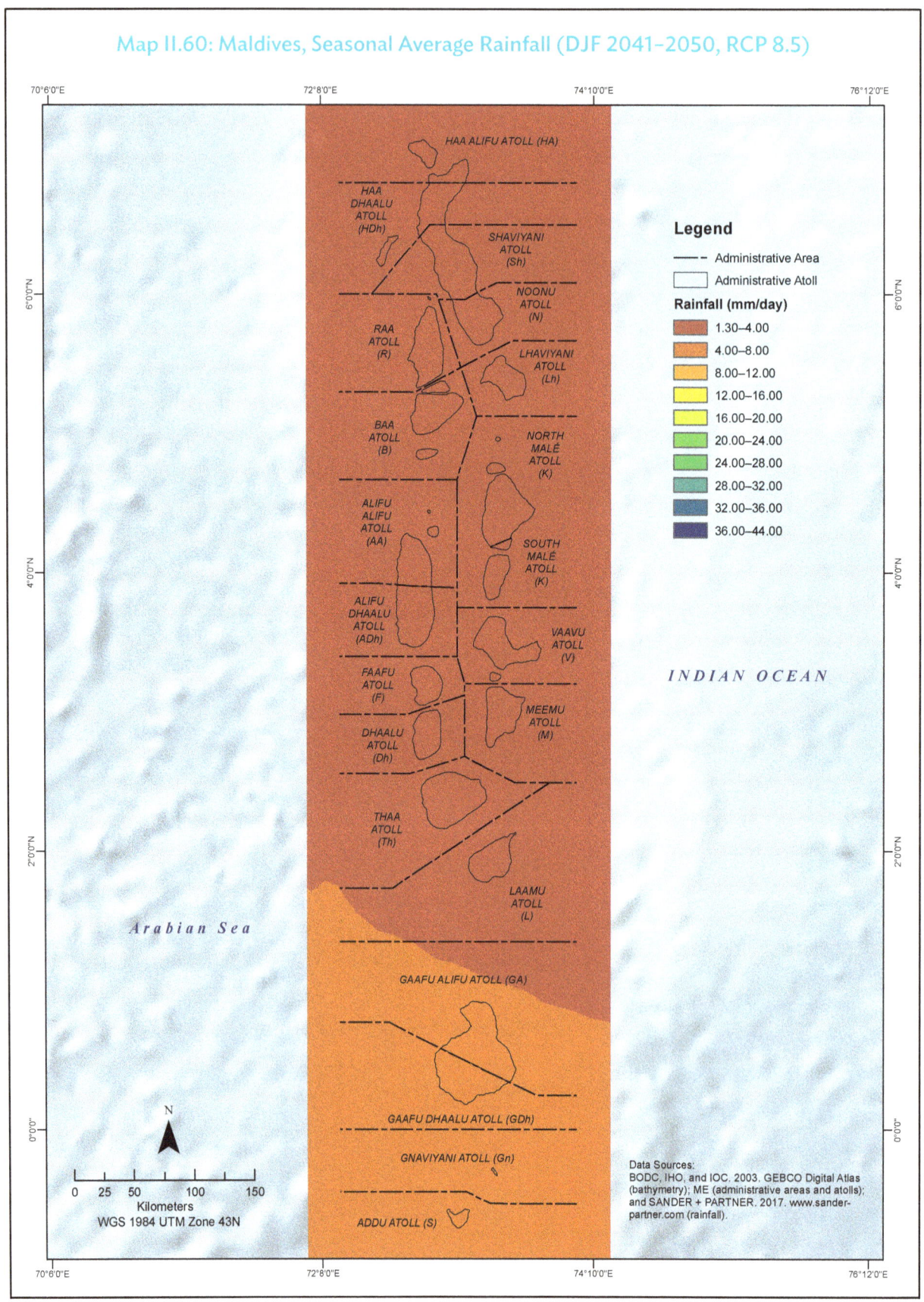
Map II.60: Maldives, Seasonal Average Rainfall (DJF 2041–2050, RCP 8.5)

Average Seasonal Rainfall Projection (MAM, RCP 8.5)

67

The MAM average seasonal rainfall projection for RCP 8.5 is similar to that of RCP 4.5. The rainfall rate for the whole Maldivian archipelago is still at its lowest range during MAM. The low rainfall rate is sustained throughout all four time periods.

- ✓ Low rainfall rate is evenly distributed across Maldives
- ✓ No visible change in rainfall across time periods
- ✓ No visible difference between RCP 8.5 and RCP 4.5 projections

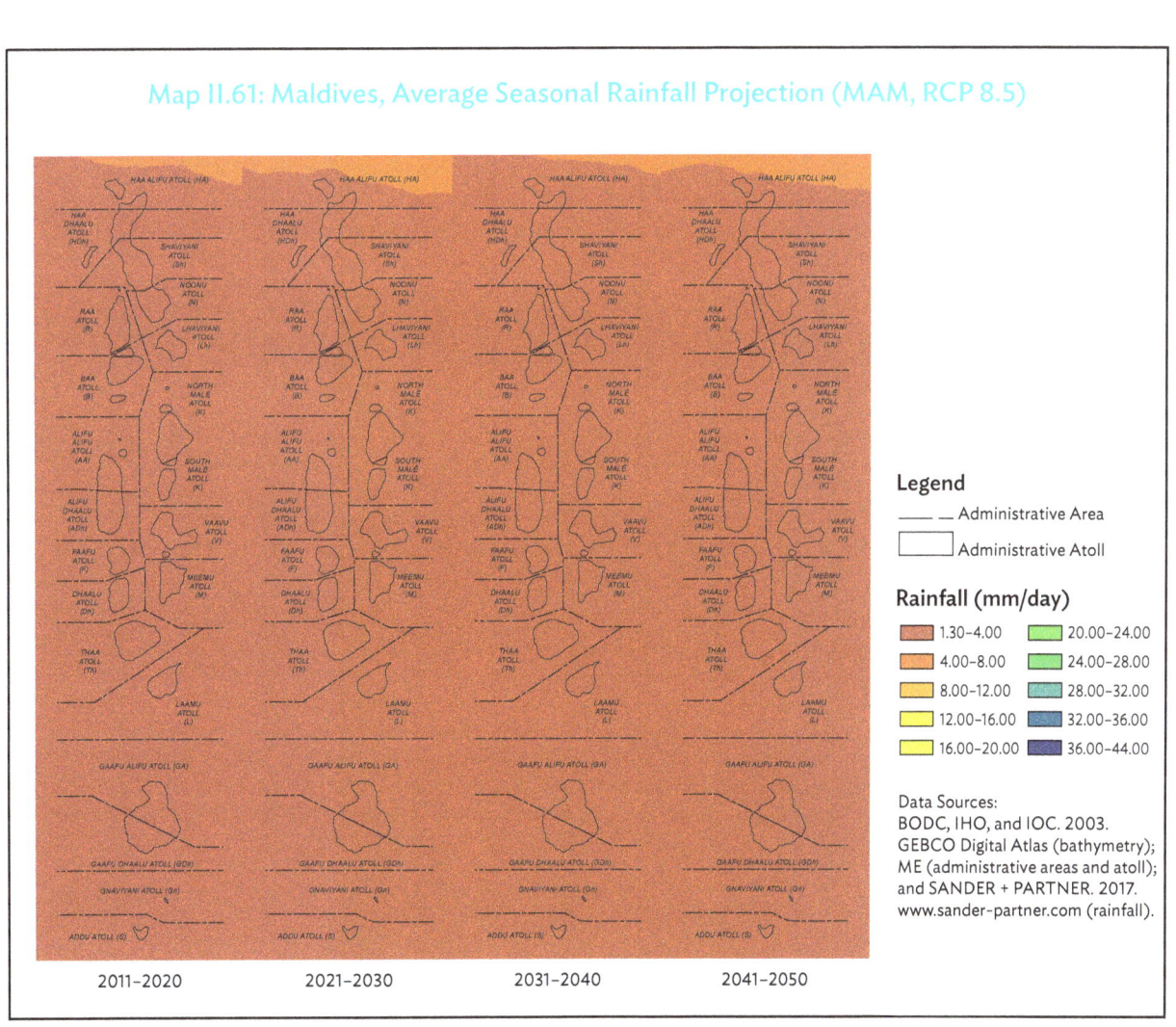

Map II.61: Maldives, Average Seasonal Rainfall Projection (MAM, RCP 8.5)

Legend
- Administrative Area
- Administrative Atoll

Rainfall (mm/day)
- 1.30–4.00
- 4.00–8.00
- 8.00–12.00
- 12.00–16.00
- 16.00–20.00
- 20.00–24.00
- 24.00–28.00
- 28.00–32.00
- 32.00–36.00
- 36.00–44.00

Data Sources:
BODC, IHO, and IOC. 2003.
GEBCO Digital Atlas (bathymetry);
ME (administrative areas and atoll);
and SANDER + PARTNER. 2017.
www.sander-partner.com (rainfall).

2011–2020 | 2021–2030 | 2031–2040 | 2041–2050

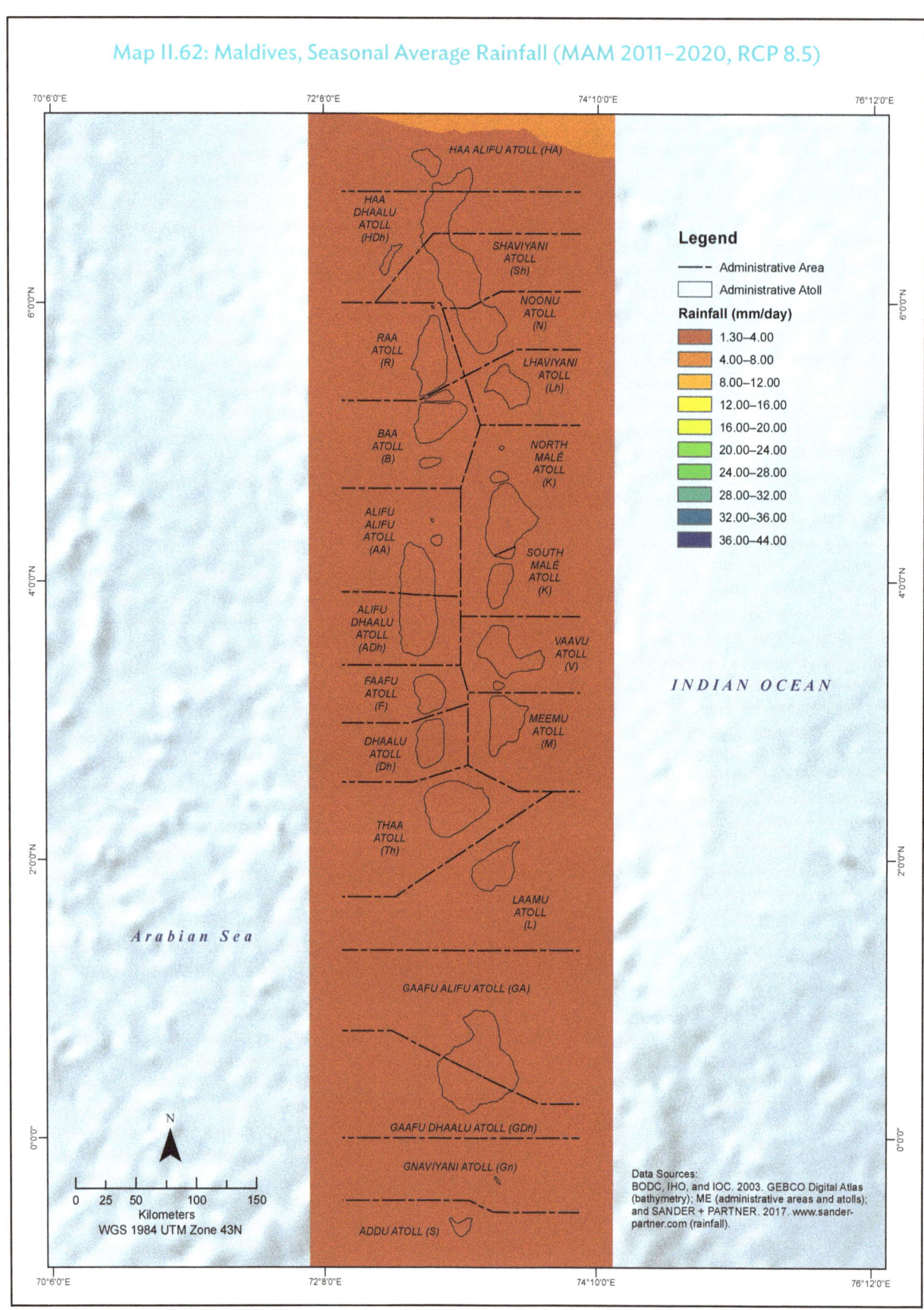

Map II.62: Maldives, Seasonal Average Rainfall (MAM 2011–2020, RCP 8.5)

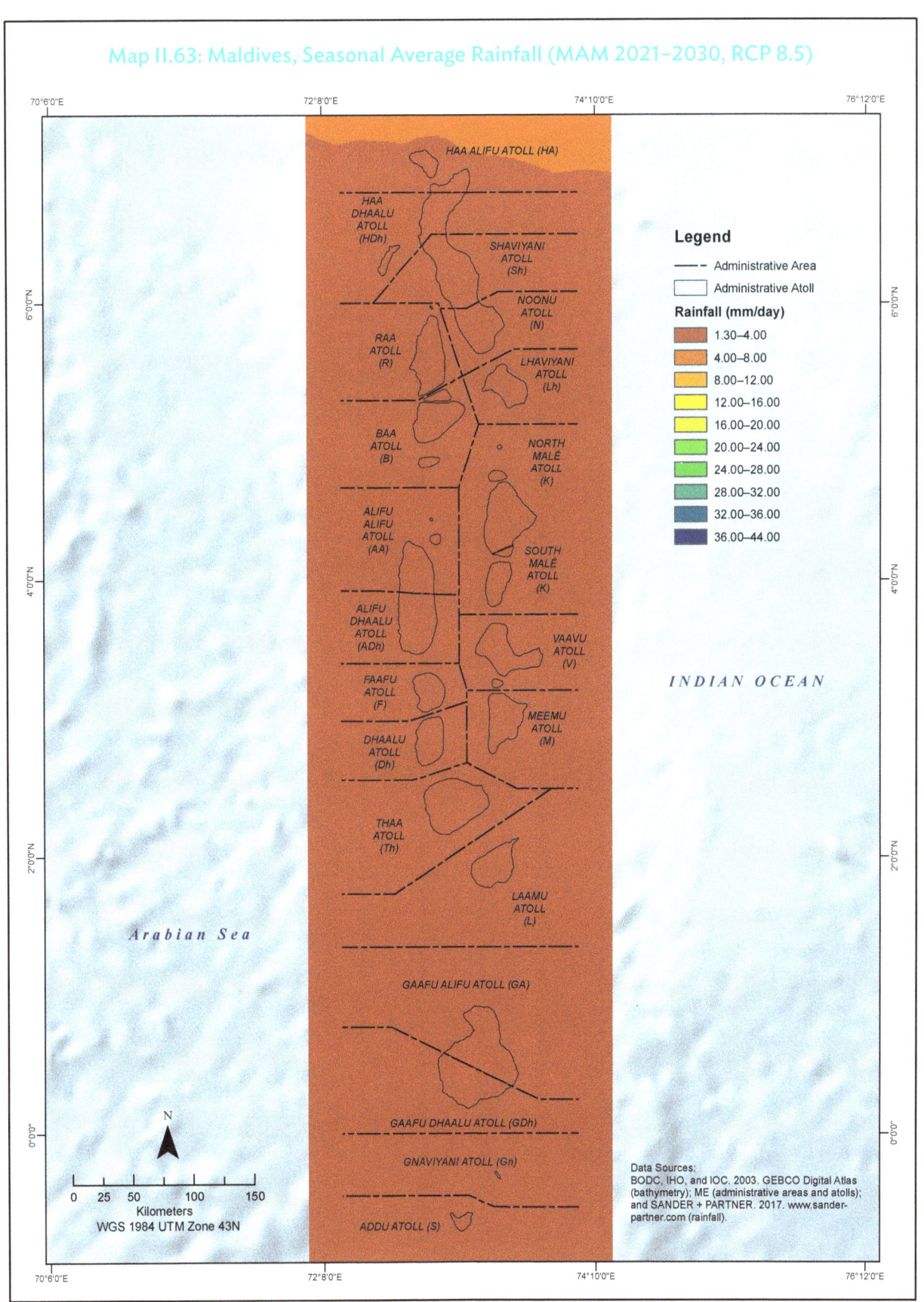

Map II.63: Maldives, Seasonal Average Rainfall (MAM 2021–2030, RCP 8.5)

70 Multihazard Risk Atlas of Maldives—Climate and Geophysical Hazards

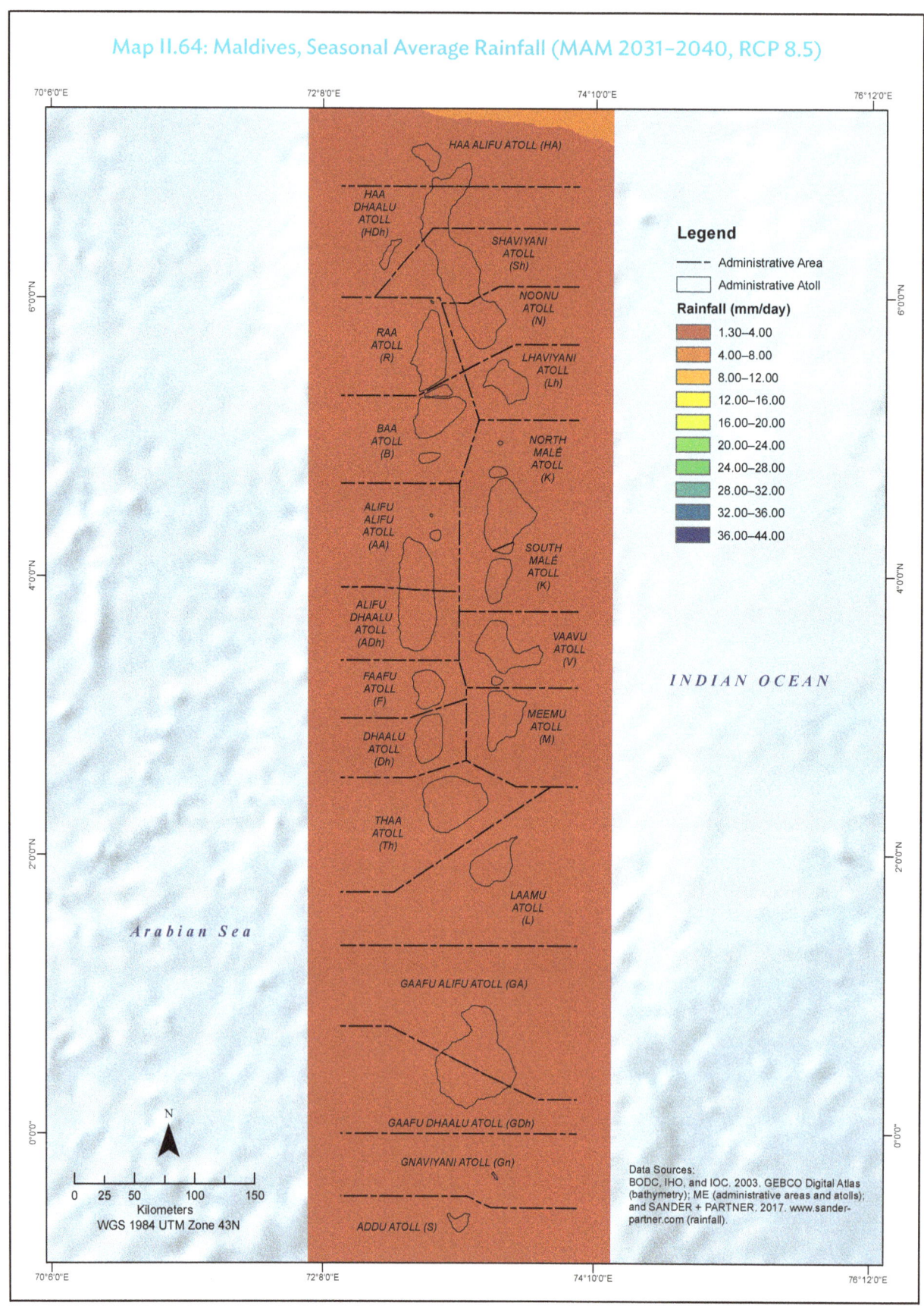

Future Climate 71

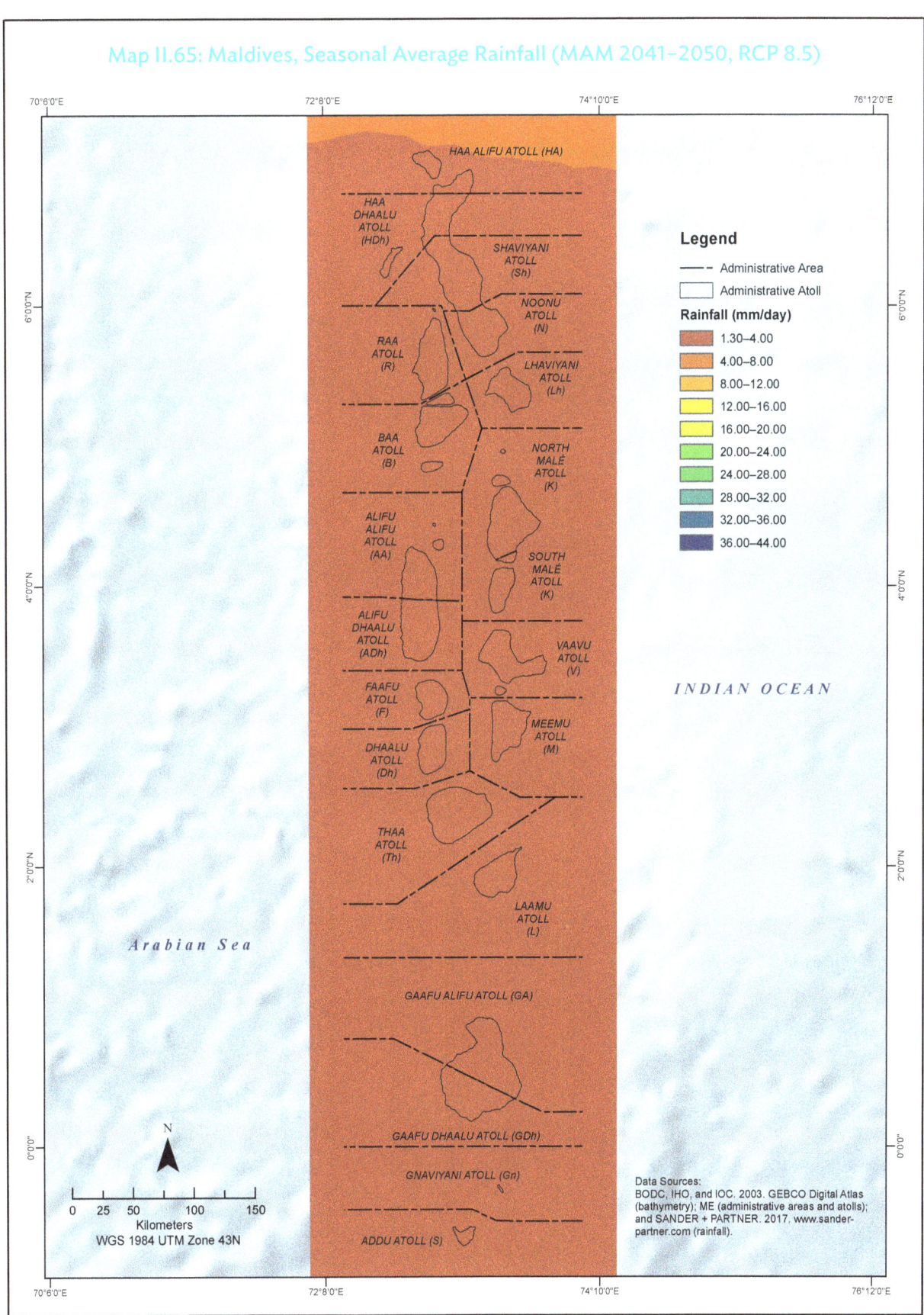

Map II.65: Maldives, Seasonal Average Rainfall (MAM 2041–2050, RCP 8.5)

Average Seasonal Rainfall Projection (JJA, RCP 8.5)

- ☑ Wetter north
- ☑ Wettest season
- ☑ Wetter north in the future
- ☑ No visible difference between RCP 8.5 and RCP 4.5 projections

JJA will continue to be the wettest months of the year, especially in the northern atolls of Maldives. Haa Alifu Atoll and Haa Dhaalu Atoll will experience the highest average daily rainfall rate (up to 12 mm/day). From the 2030s to the 2040s, Shaviyani Atoll will also experience this average daily rainfall rate.

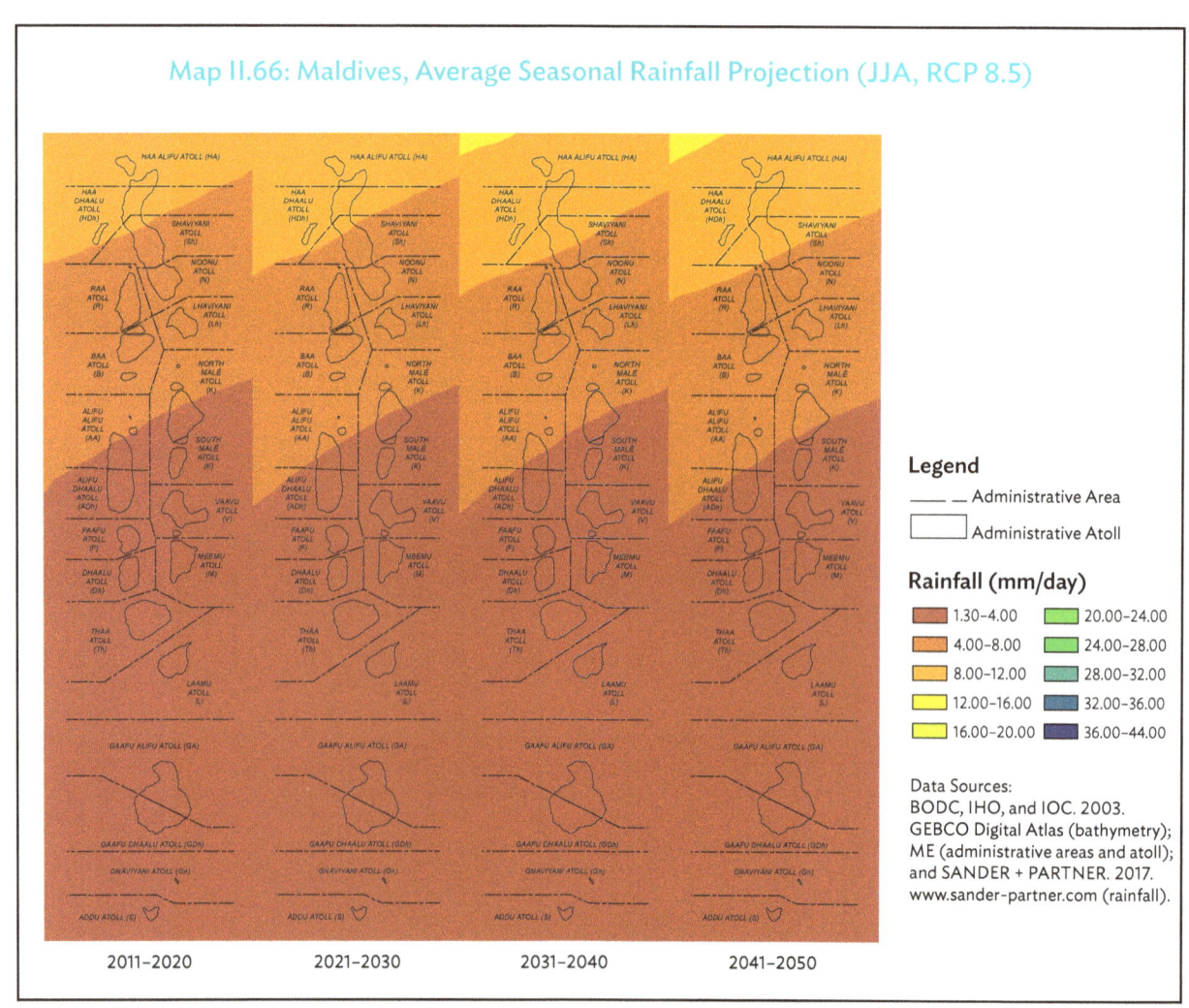

Map II.66: Maldives, Average Seasonal Rainfall Projection (JJA, RCP 8.5)

Future Climate 73

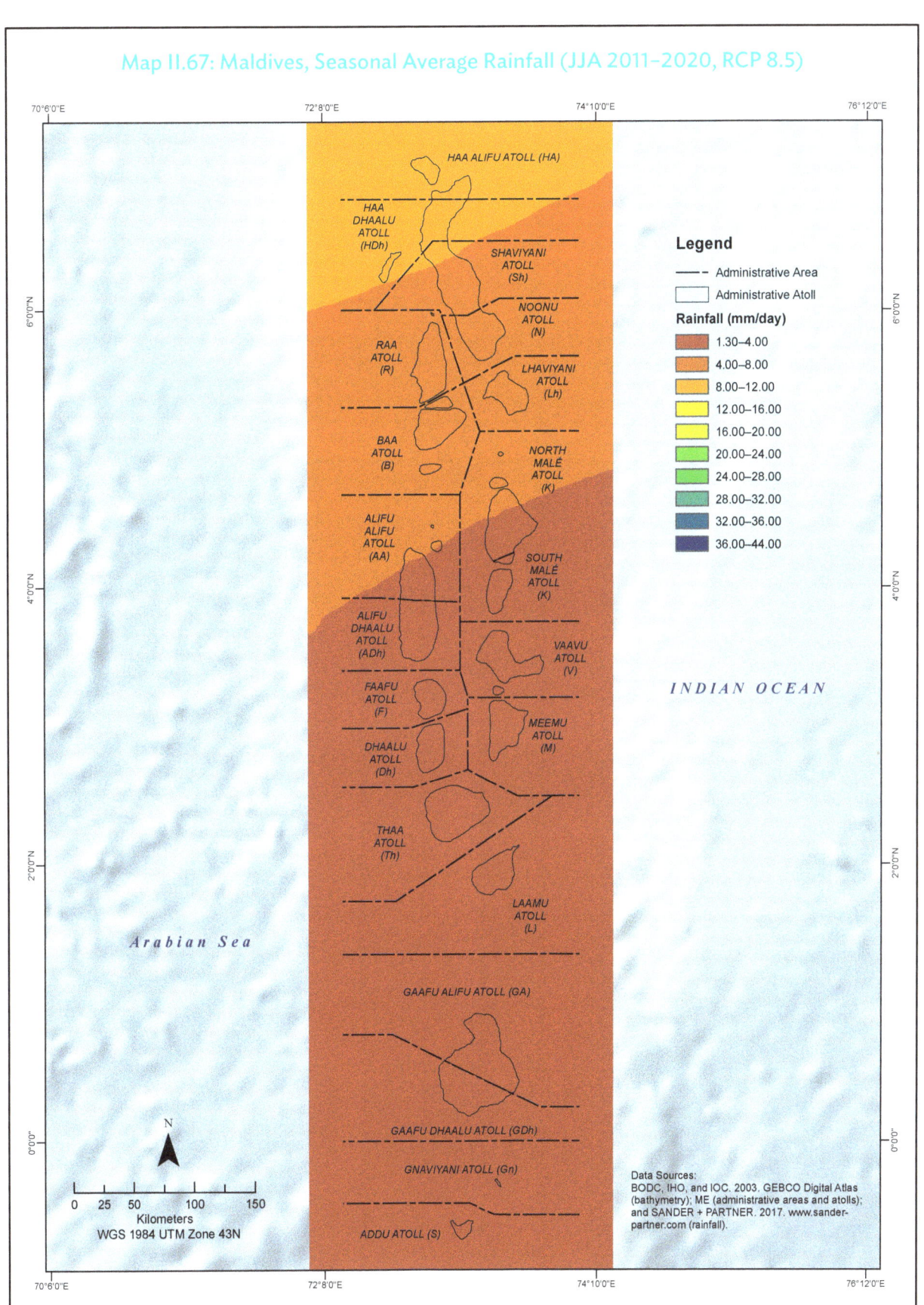

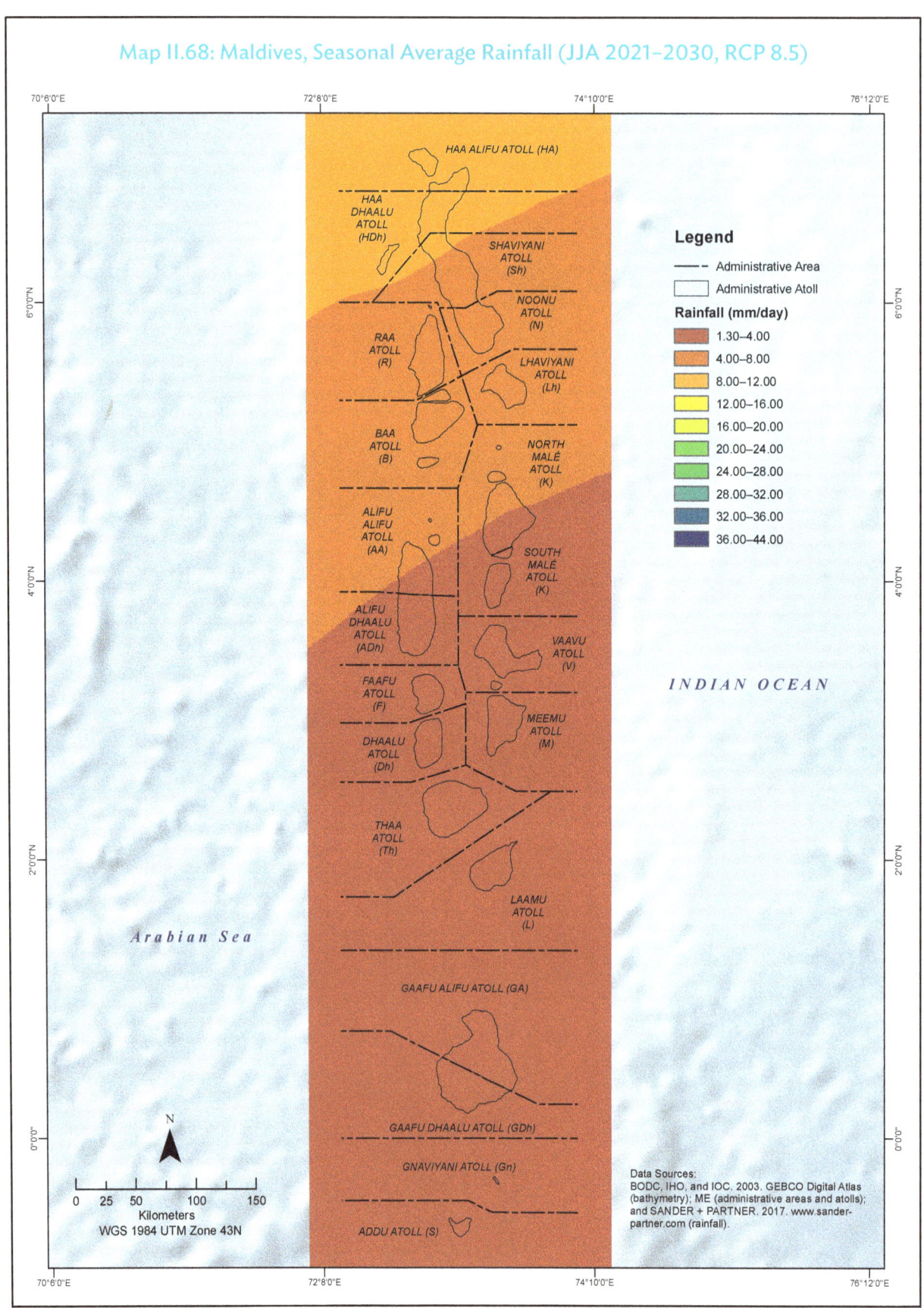

Map II.68: Maldives, Seasonal Average Rainfall (JJA 2021–2030, RCP 8.5)

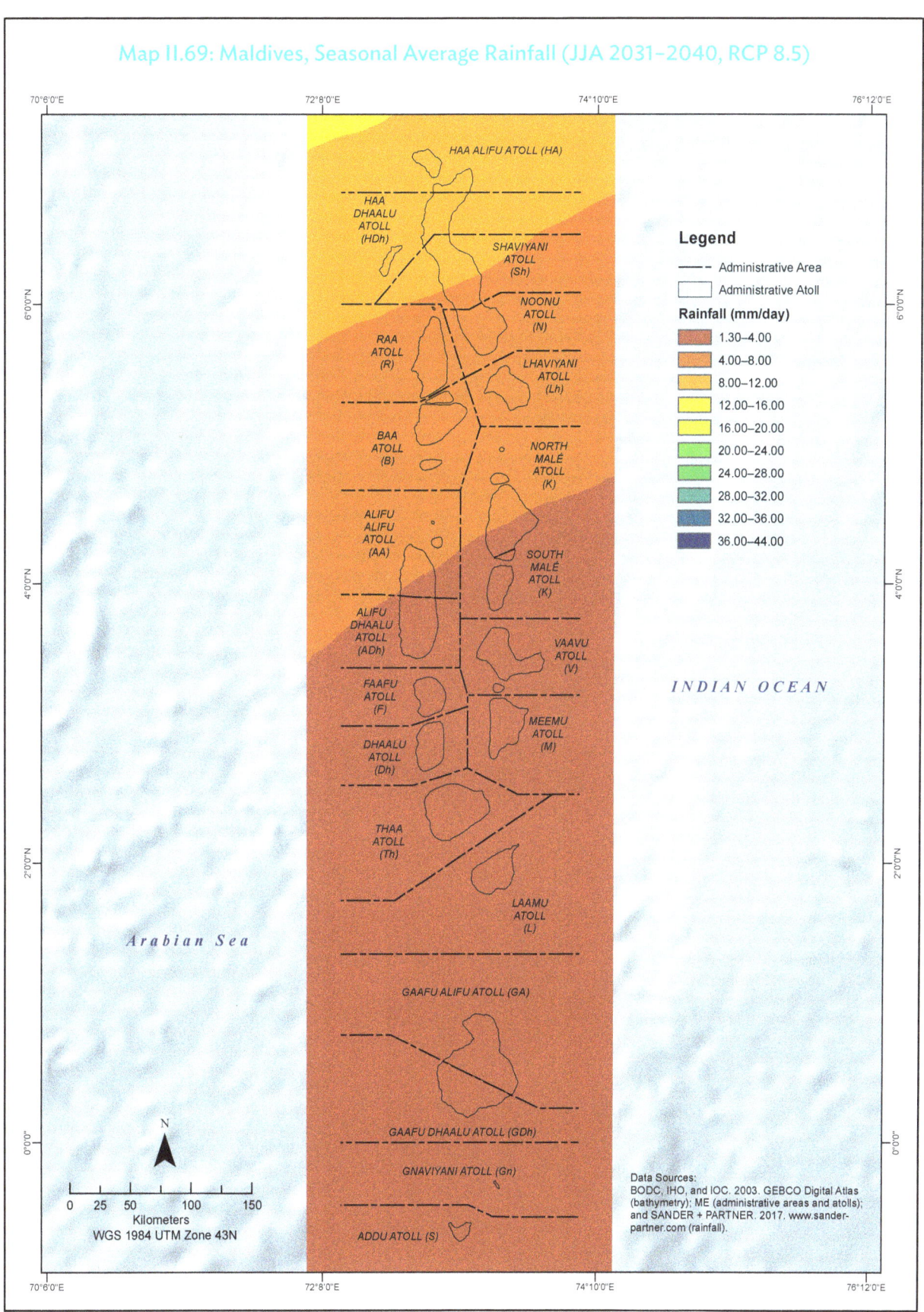

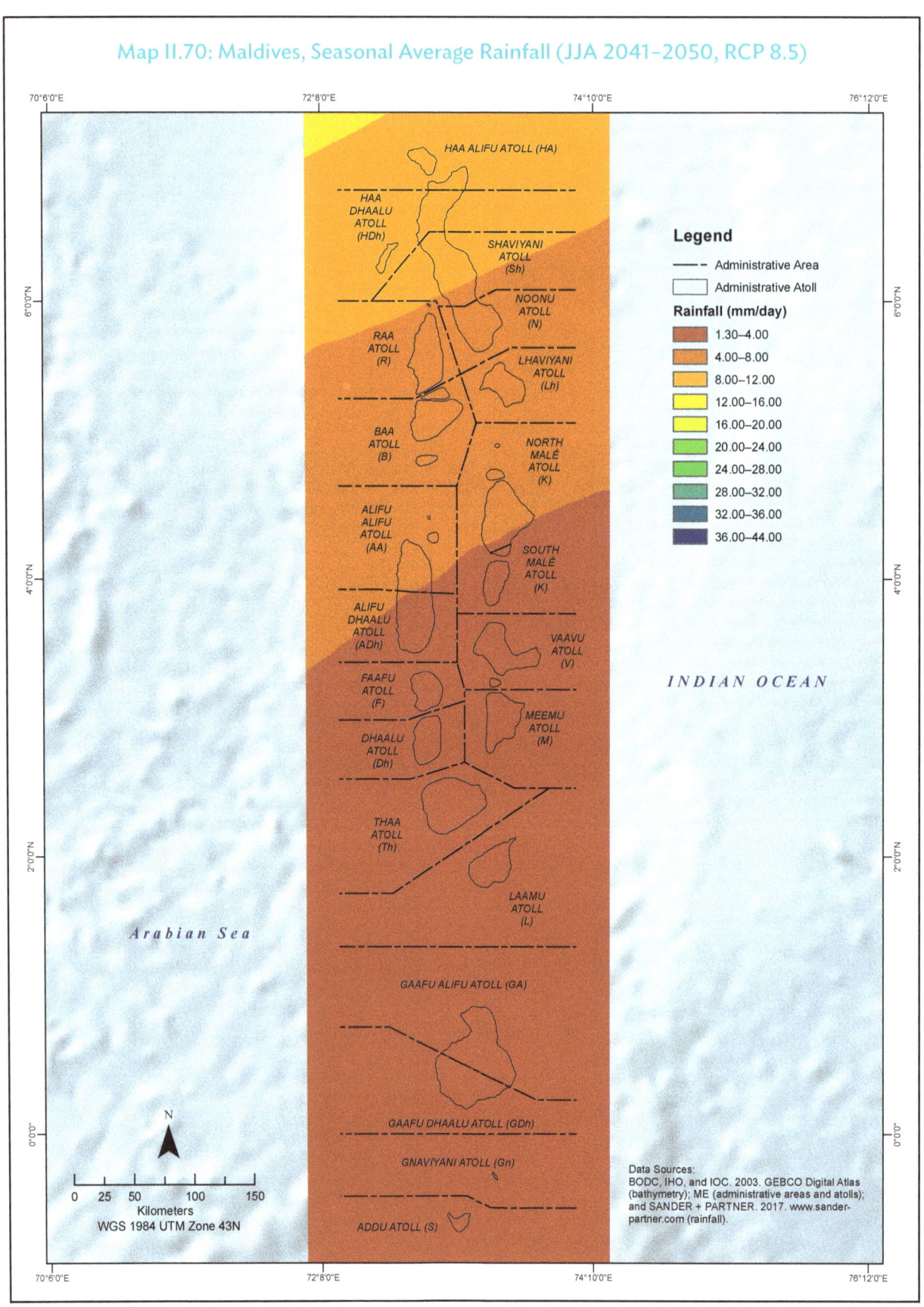

Map II.70: Maldives, Seasonal Average Rainfall (JJA 2041–2050, RCP 8.5)

Average Seasonal Rainfall Projection (SON, RCP 8.5)

77

The time series of projected average rainfall during SON based on RCP 8.5 shows an increasing area covered by a rainfall rate of 4–8 mm/day. This translates to increasing rainfall in the middle portion of Maldives in the 2020s. During SON, the northern half of Maldives has a rainfall rate of 4–8 mm/day and the southern half has a low rainfall rate of 1.3–4 mm/day.

- ☑ Wetter north
- ☑ More northern atolls with higher rainfall rate (4–8 mm/day) in the future
- ☑ Greater area with 4–8 mm/day rainfall rate in the future in RCP 8.5 compared with RCP 4.5

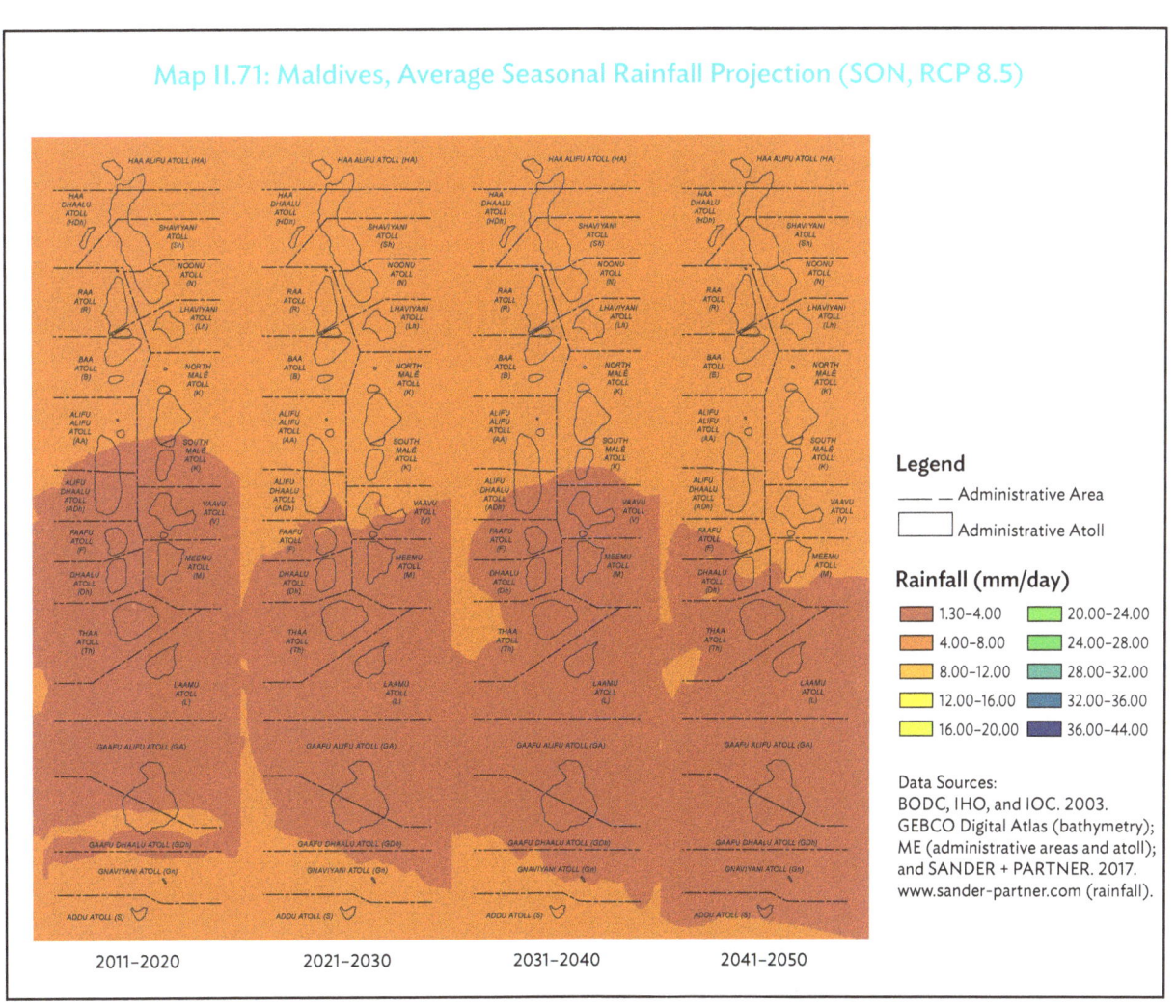

Map II.71: Maldives, Average Seasonal Rainfall Projection (SON, RCP 8.5)

Legend
- Administrative Area
- Administrative Atoll

Rainfall (mm/day)
- 1.30–4.00
- 4.00–8.00
- 8.00–12.00
- 12.00–16.00
- 16.00–20.00
- 20.00–24.00
- 24.00–28.00
- 28.00–32.00
- 32.00–36.00
- 36.00–44.00

Data Sources:
BODC, IHO, and IOC. 2003.
GEBCO Digital Atlas (bathymetry);
ME (administrative areas and atoll);
and SANDER + PARTNER. 2017.
www.sander-partner.com (rainfall).

2011–2020 | 2021–2030 | 2031–2040 | 2041–2050

78 Multihazard Risk Atlas of Maldives—Climate and Geophysical Hazards

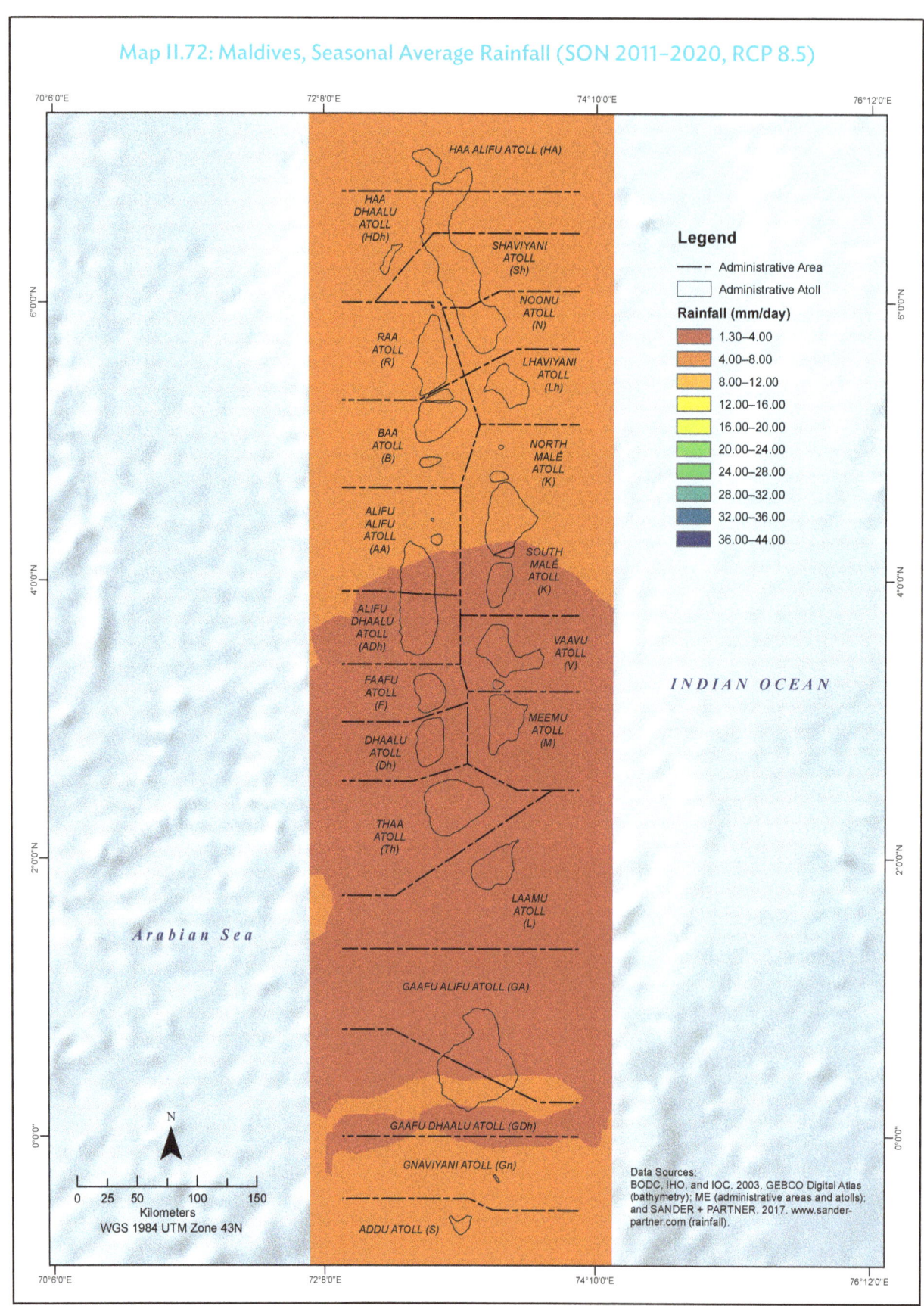

Future Climate

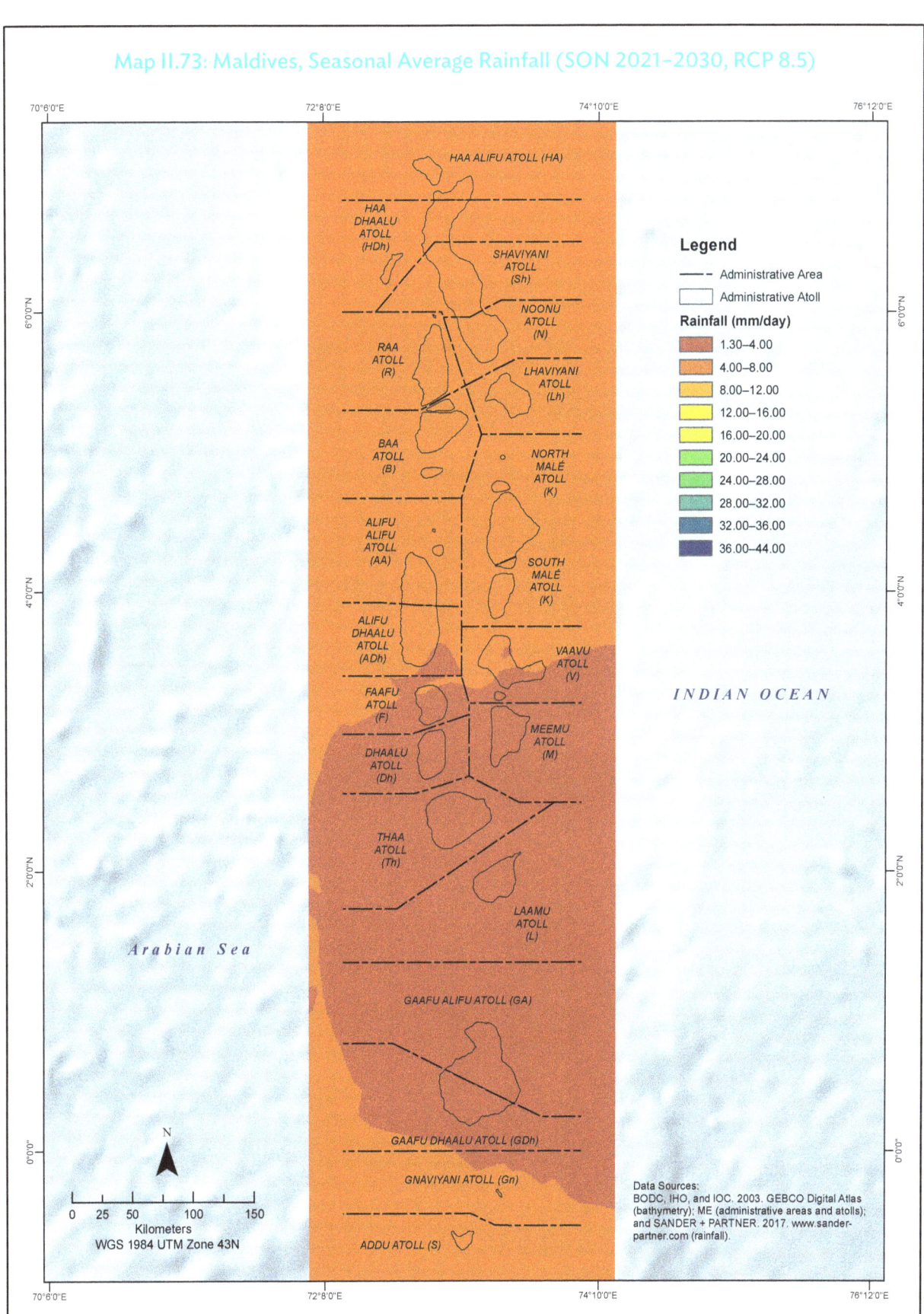

Map II.73: Maldives, Seasonal Average Rainfall (SON 2021–2030, RCP 8.5)

80 Multihazard Risk Atlas of Maldives—Climate and Geophysical Hazards

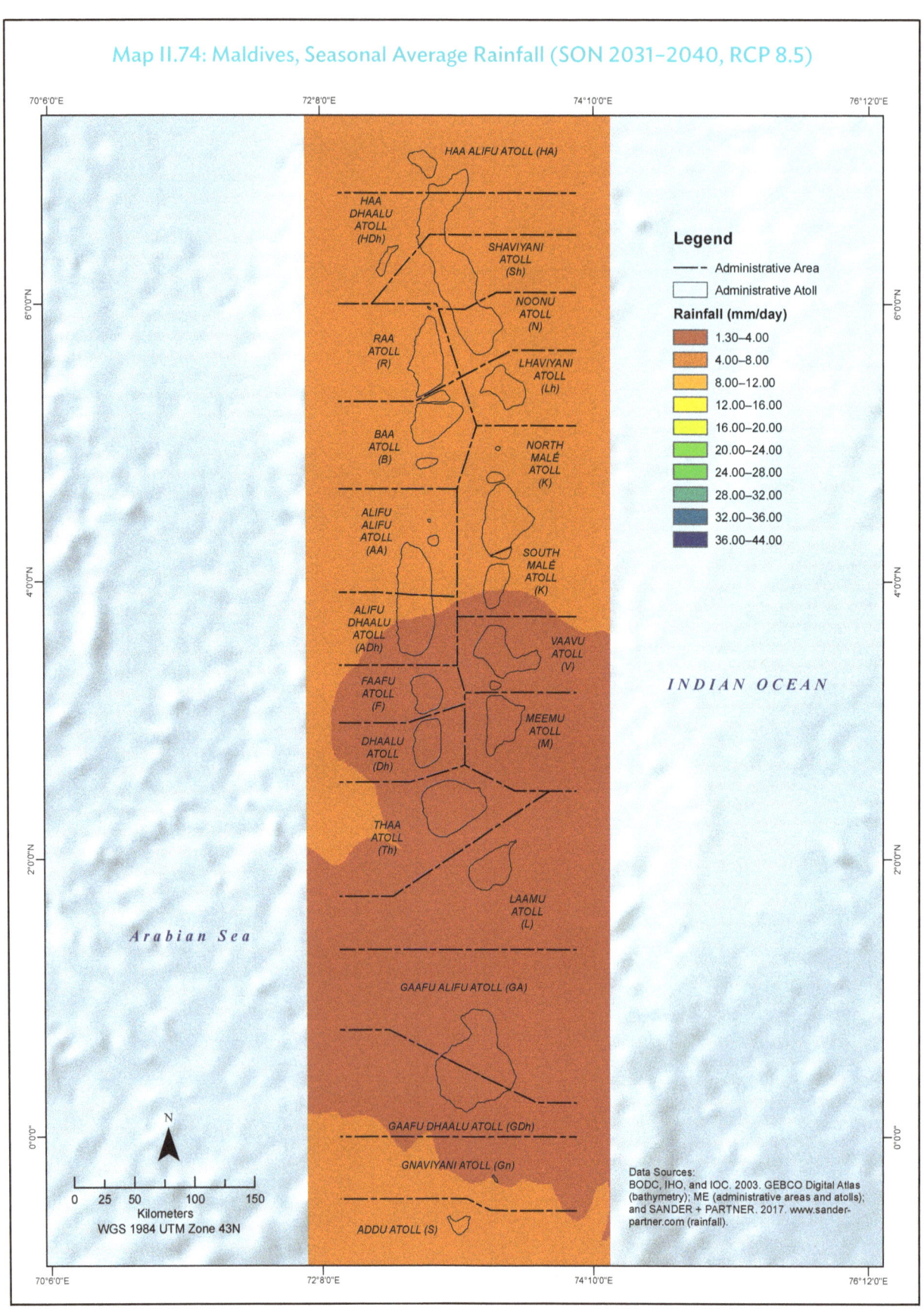

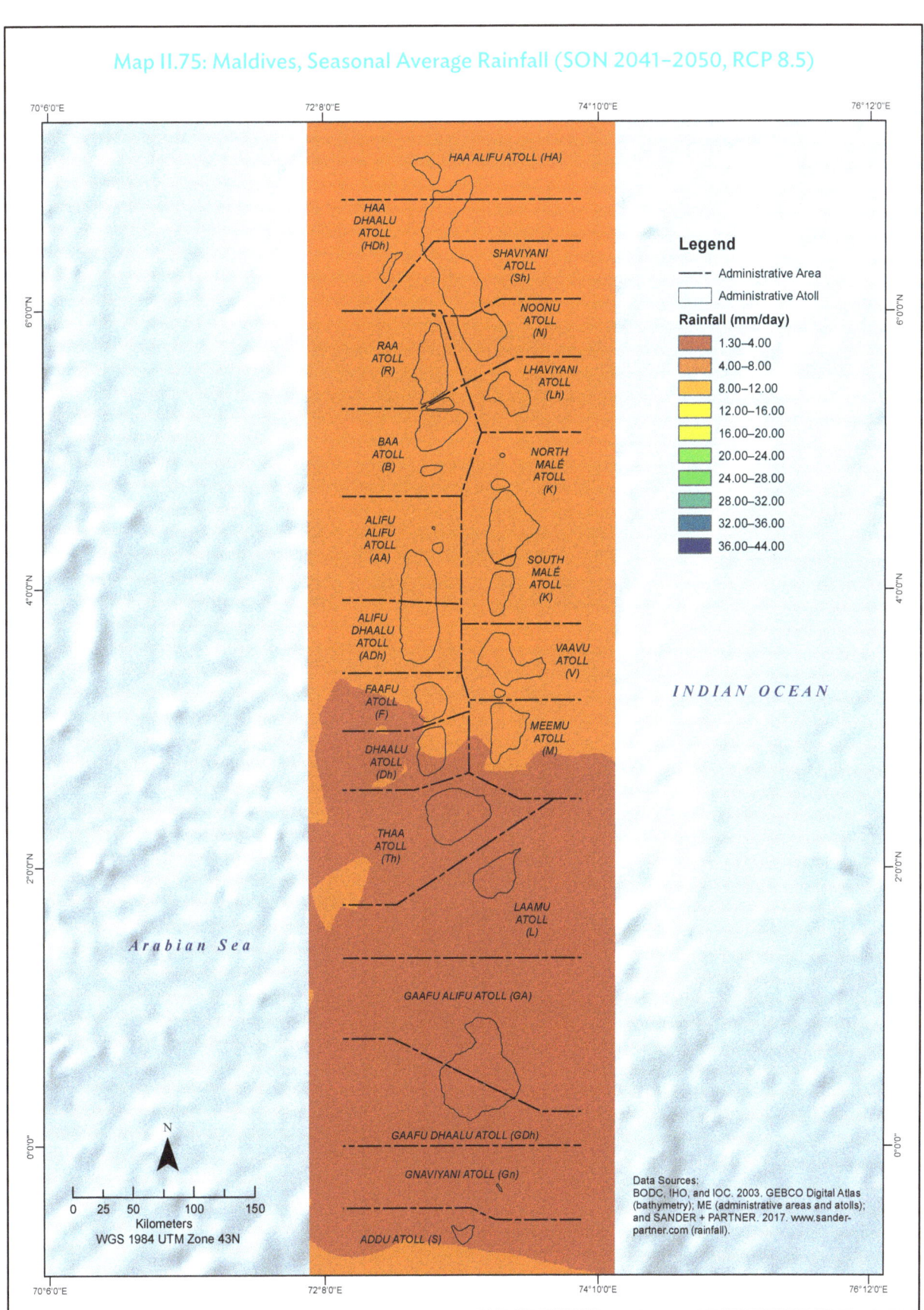

Average Annual Temperature Projection (RCP 4.5)

- [✓] Warmer north
- [✓] Warmer Maldives in the future

In this decade, the average annual temperature in Maldives (with the exception of Haafu Alifu) is 28.4°C–28.8°C. Maldives will have warmer days (average of 29.05°C–30.15°C) in the 2030s and 2040s, particularly in northern Maldives where the average annual temperature will range from 29.60°C to 30.15°C.

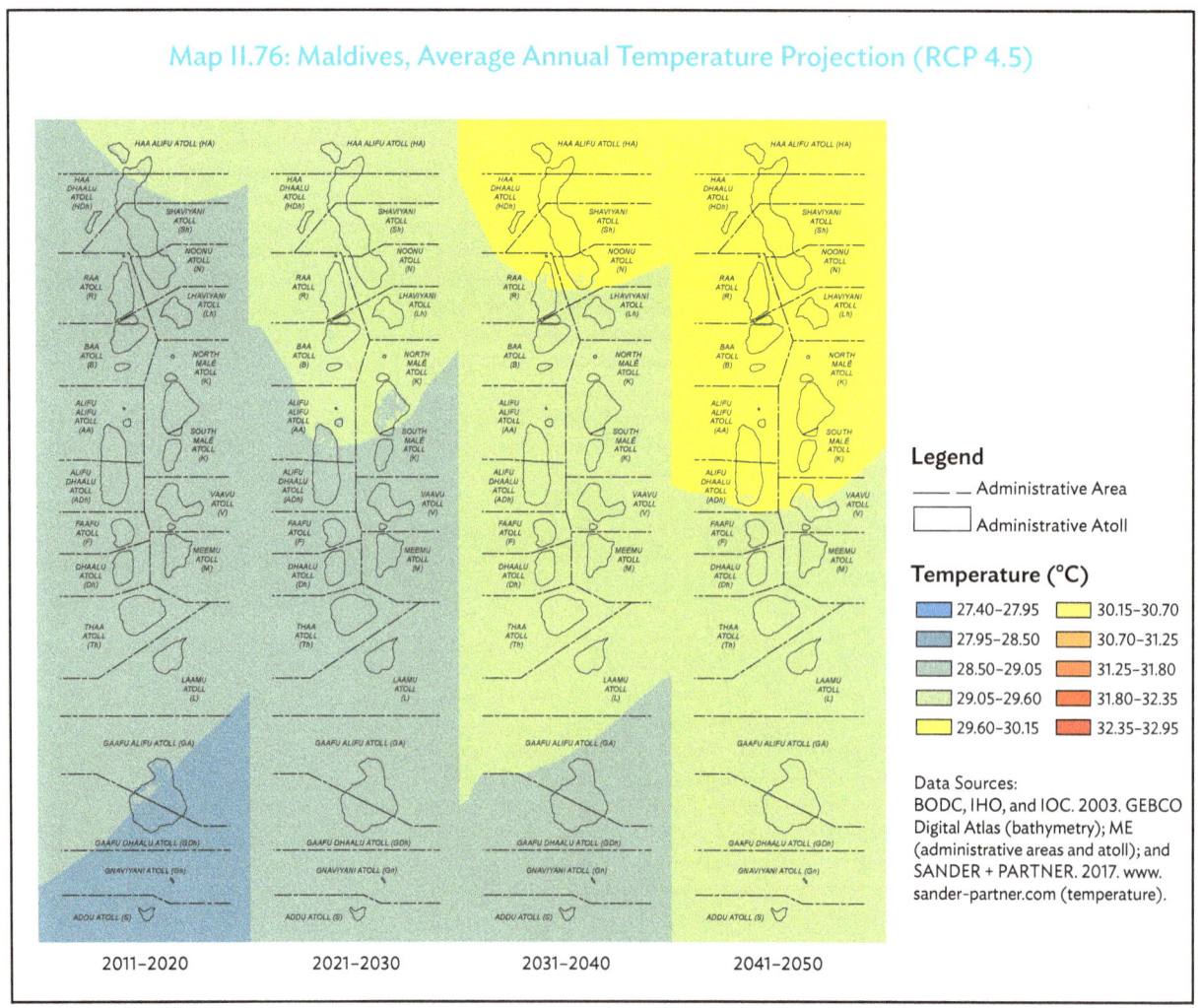

Map II.76: Maldives, Average Annual Temperature Projection (RCP 4.5)

Future Climate 83

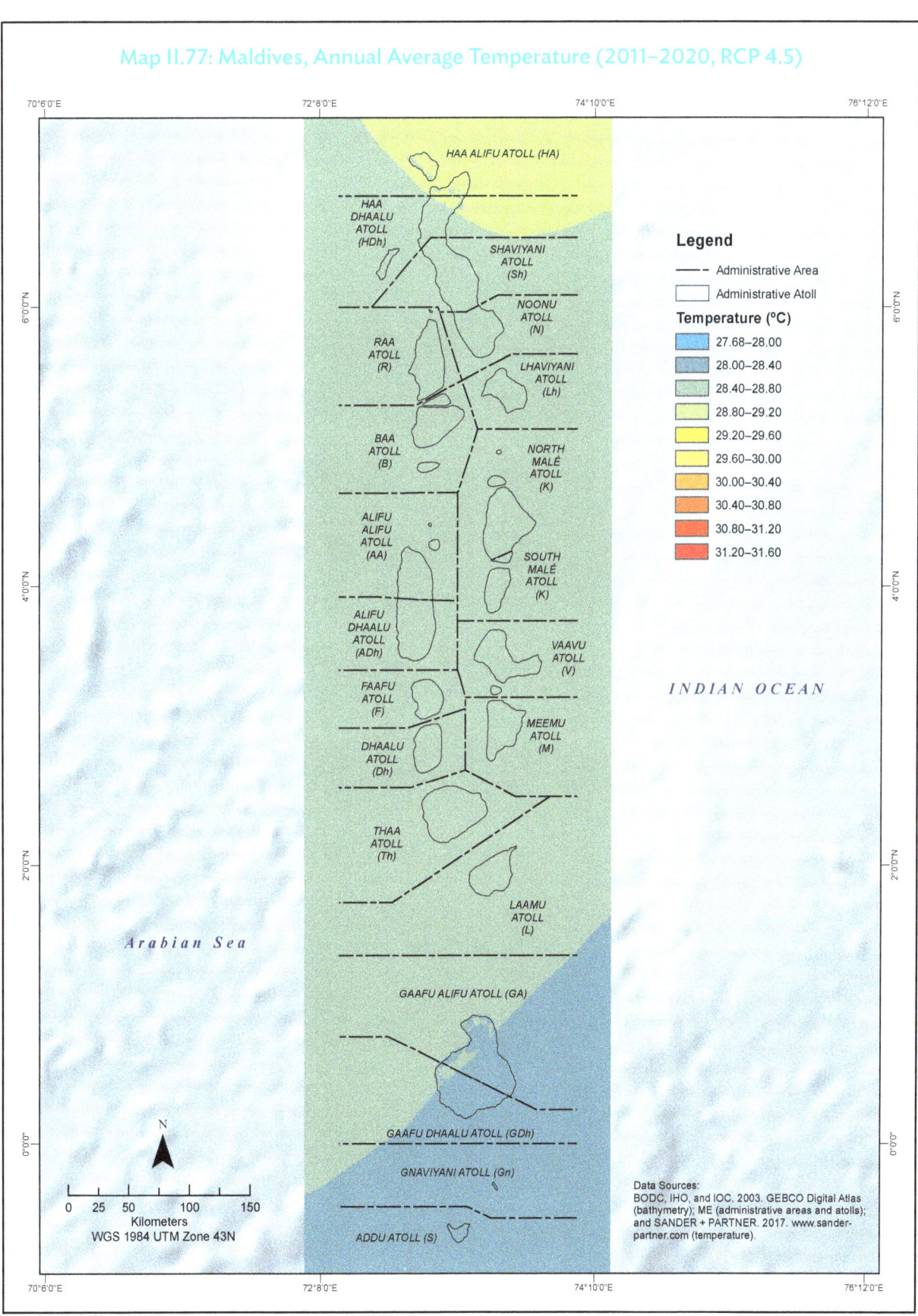

Map II.77: Maldives, Annual Average Temperature (2011–2020, RCP 4.5)

84 Multihazard Risk Atlas of Maldives—Climate and Geophysical Hazards

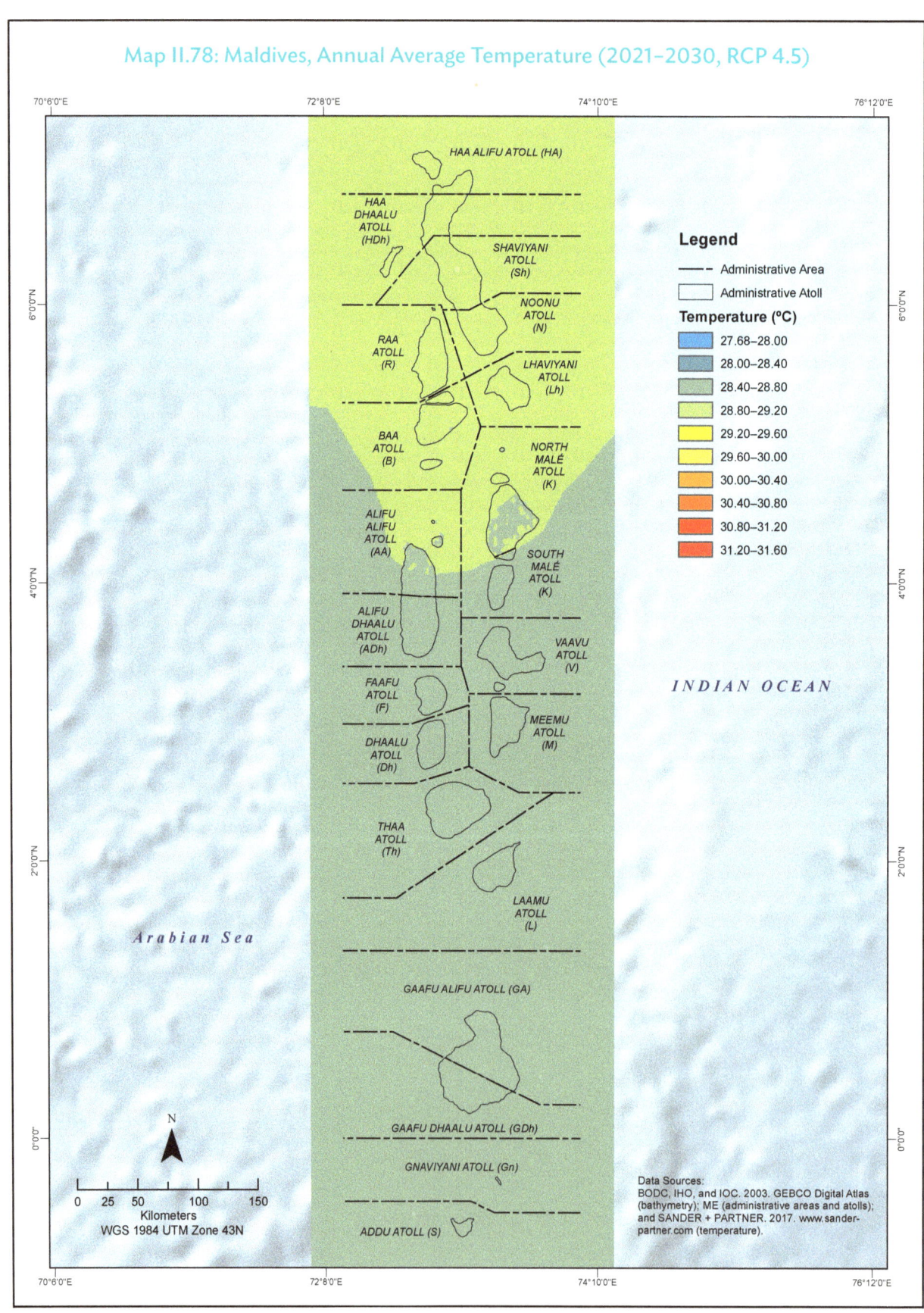

Map II.78: Maldives, Annual Average Temperature (2021–2030, RCP 4.5)

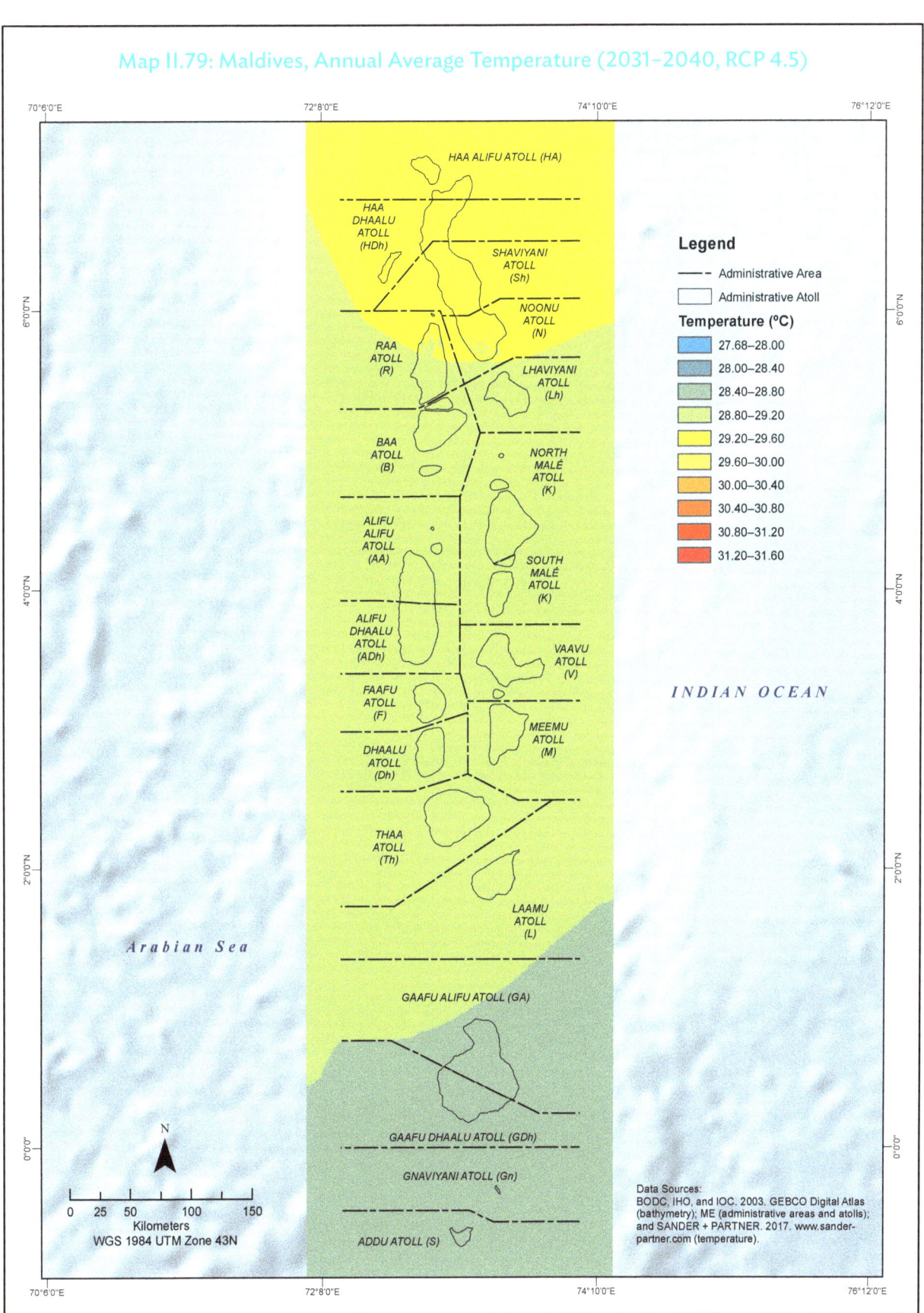
Map II.79: Maldives, Annual Average Temperature (2031–2040, RCP 4.5)

86　Multihazard Risk Atlas of Maldives—Climate and Geophysical Hazards

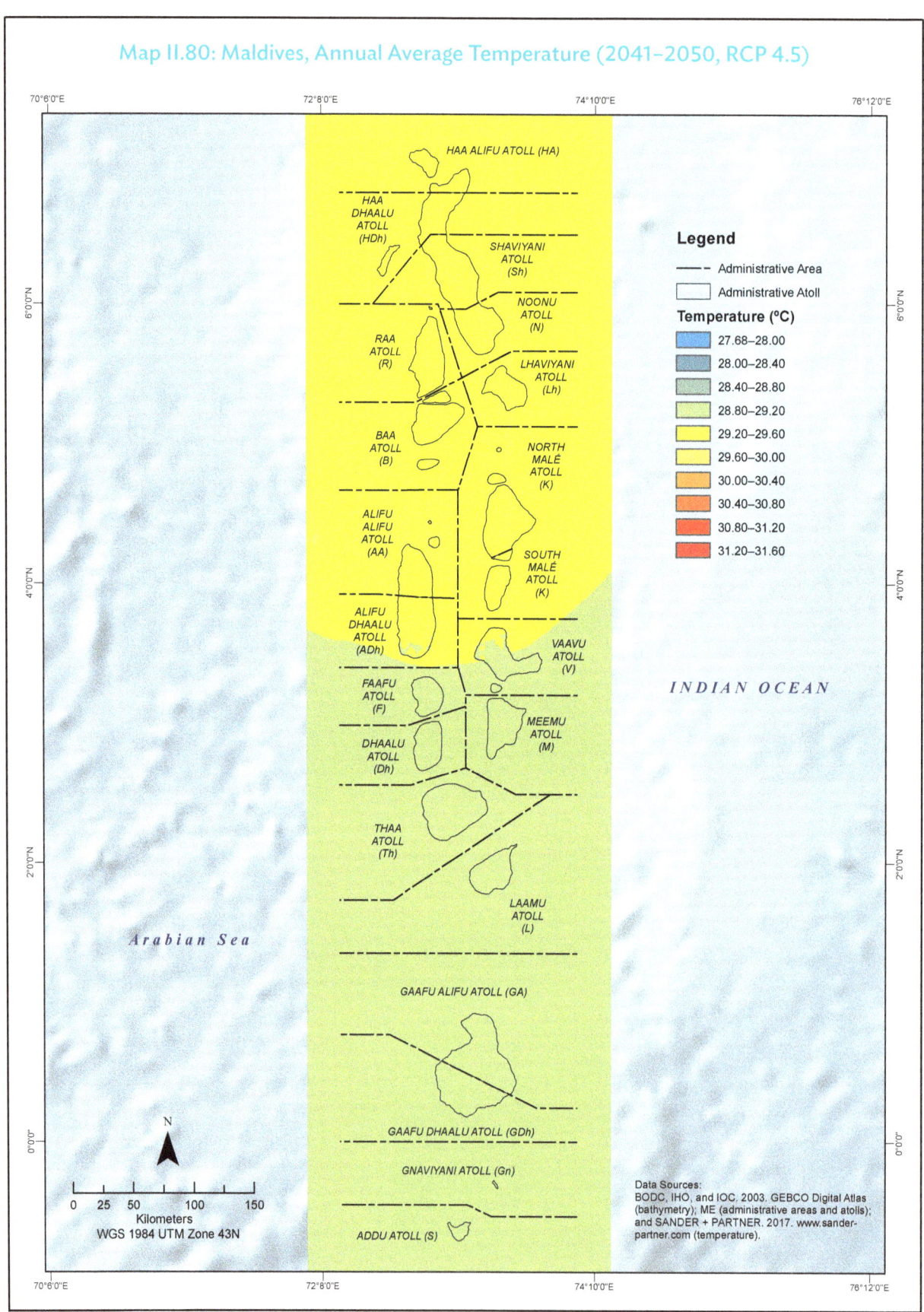

Average Seasonal Temperature Projection (DJF, RCP 4.5)

87

The current average temperature during DJF in Maldives ranges from 27.95°C to 28.50°C, with higher temperature in the northeastern region. In the 2030s and 2040s, Maldives will have warmer days during DJF. People living in the north will experience an average temperature ranging from 29.05°C to 29.60°C beginning 2031. By 2050, this average temperature during DJF will be experienced across the country.

- ✓ Warmer north
- ✓ Warmer Maldives in the future

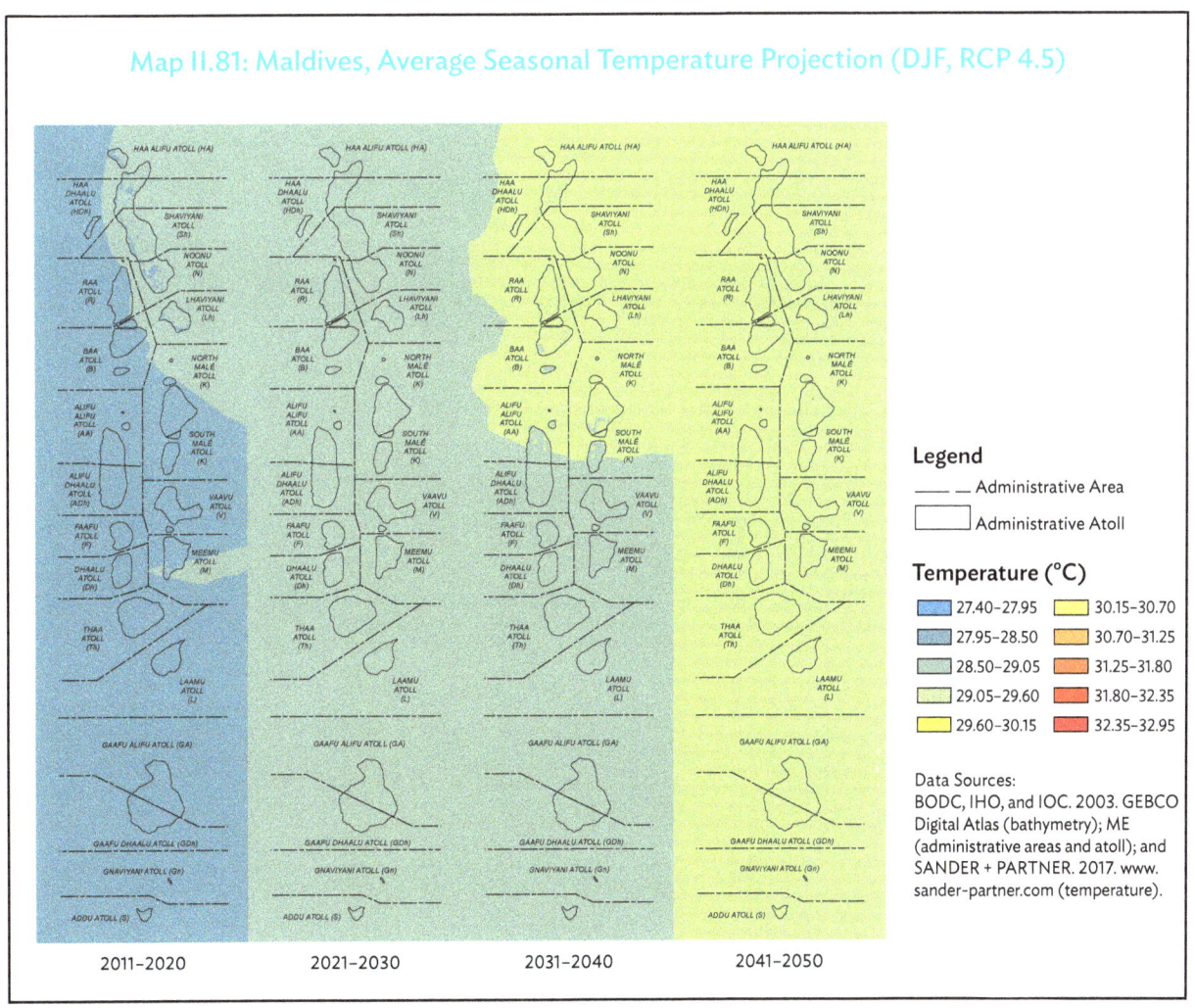

Map II.81: Maldives, Average Seasonal Temperature Projection (DJF, RCP 4.5)

Data Sources: BODC, IHO, and IOC. 2003. GEBCO Digital Atlas (bathymetry); ME (administrative areas and atoll); and SANDER + PARTNER. 2017. www.sander-partner.com (temperature).

88 Multihazard Risk Atlas of Maldives—Climate and Geophysical Hazards

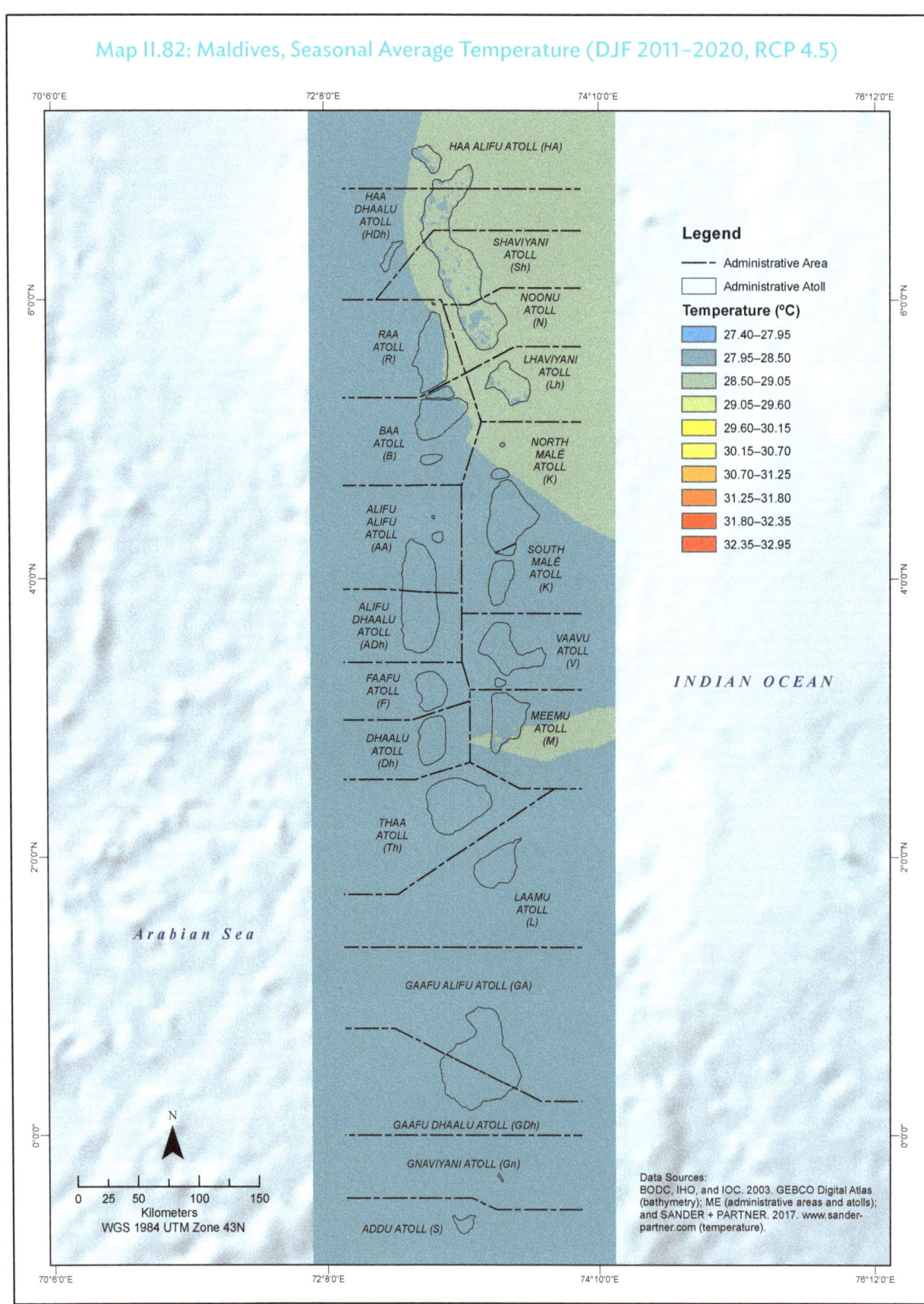

Map II.82: Maldives, Seasonal Average Temperature (DJF 2011–2020, RCP 4.5)

Future Climate

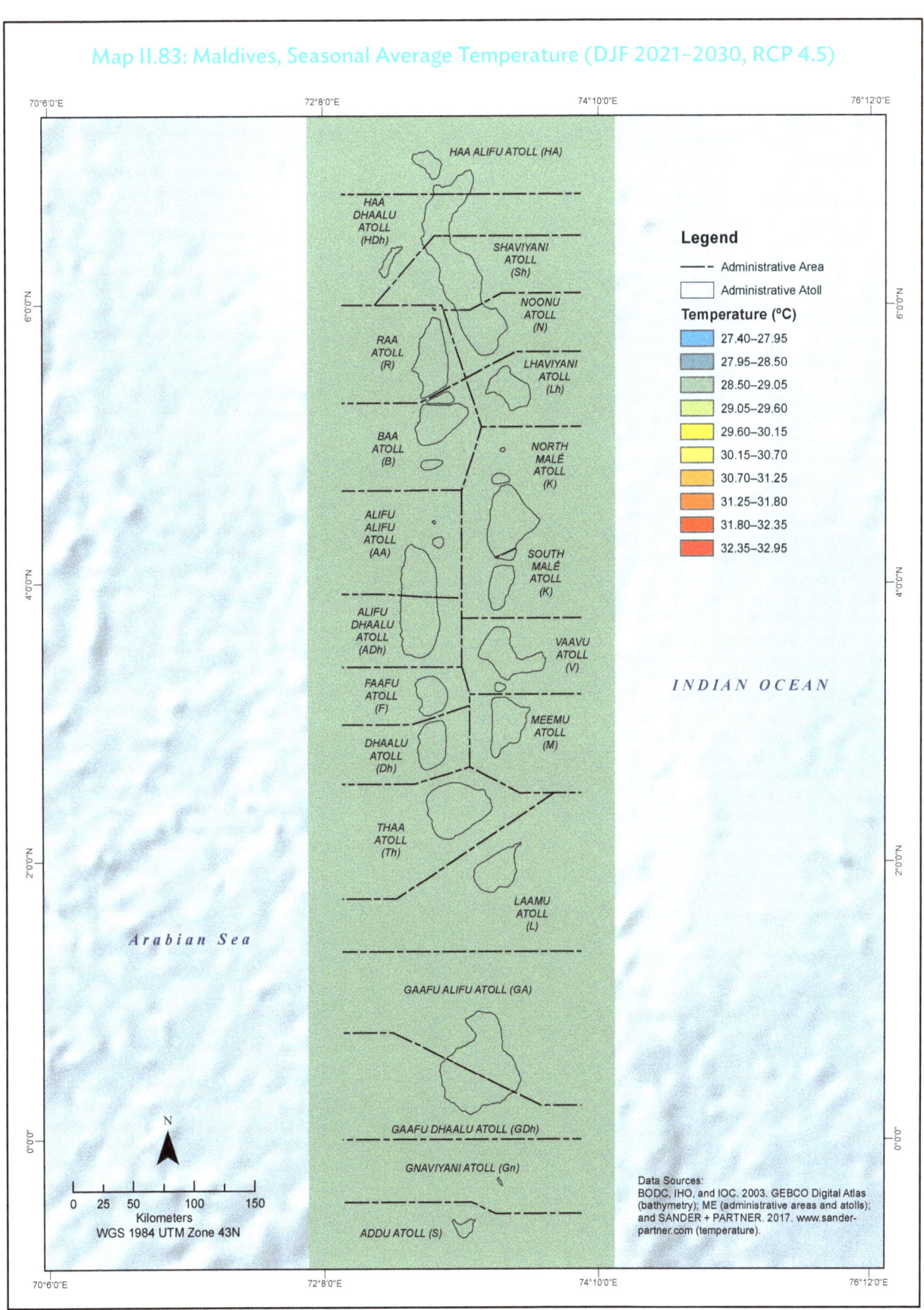

Map II.83: Maldives, Seasonal Average Temperature (DJF 2021–2030, RCP 4.5)

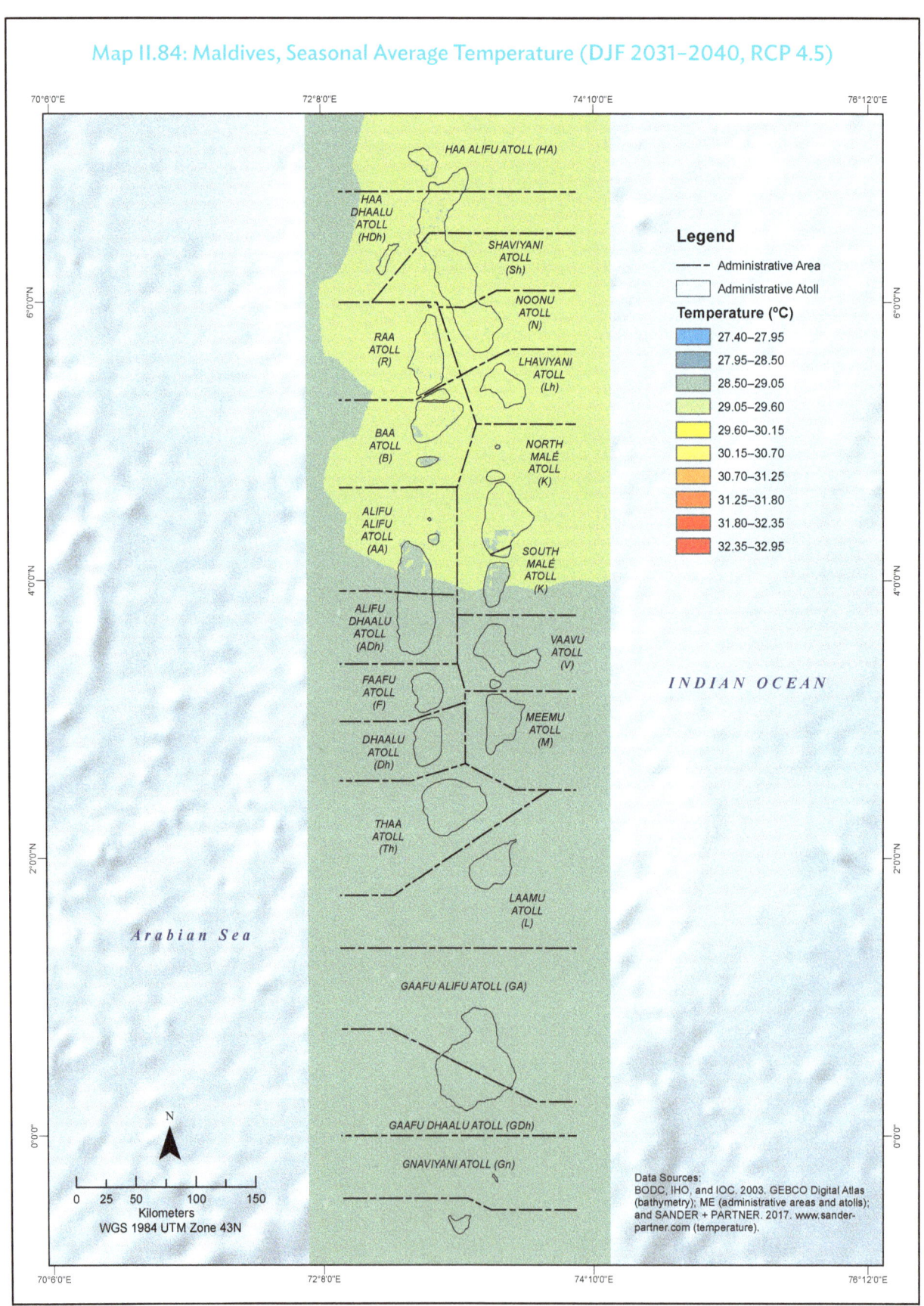

Map II.84: Maldives, Seasonal Average Temperature (DJF 2031–2040, RCP 4.5)

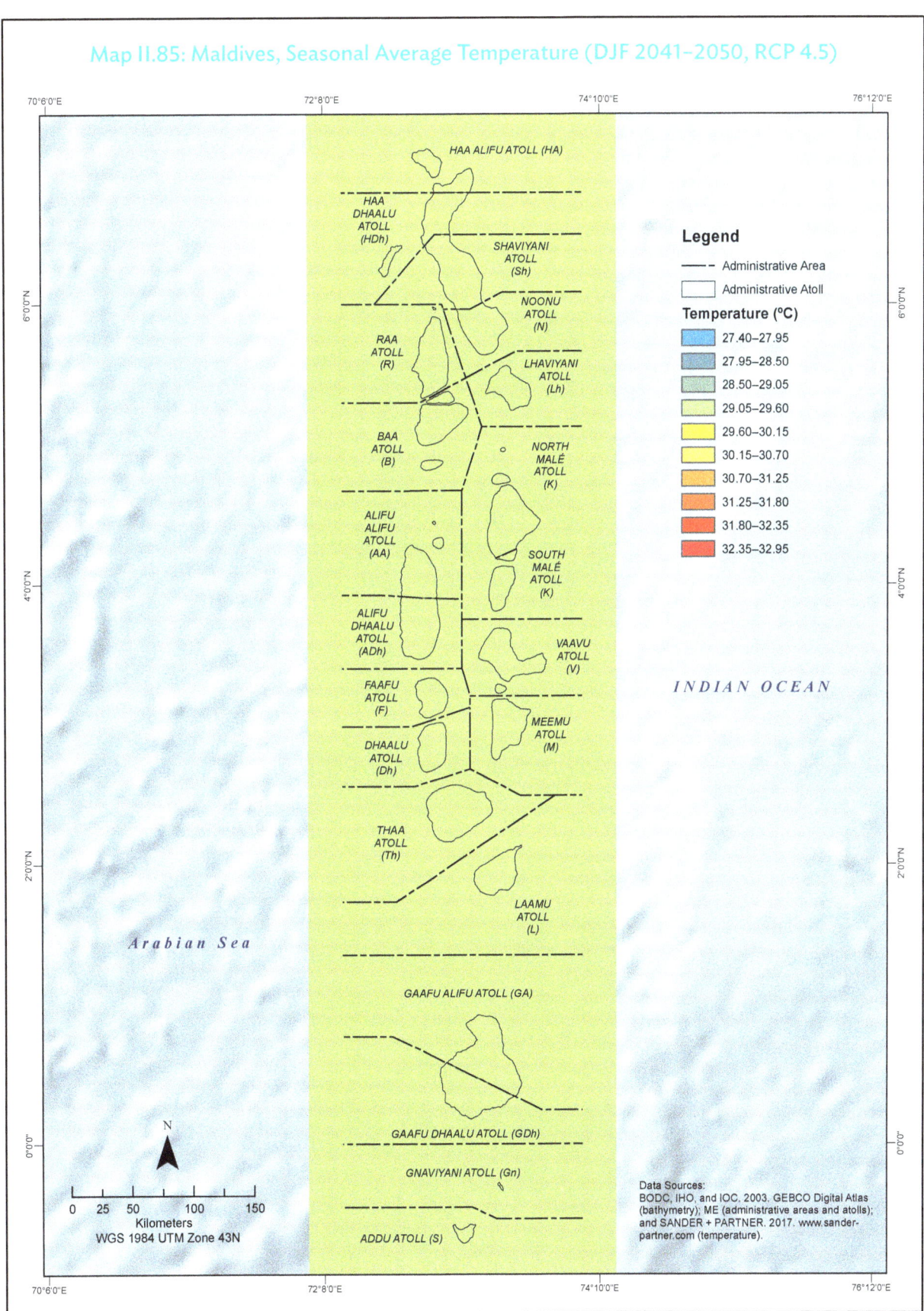
Map II.85: Maldives, Seasonal Average Temperature (DJF 2041–2050, RCP 4.5)

Average Seasonal Temperature Projection (MAM, RCP 4.5)

- ☑ Warmer north
- ☑ Warmer Maldives in the future
- ☑ Warmest months are MAM

MAM are the warmest months, especially in northern Maldives. The average temperature ranges from 28.50°C to 29.05°C in the 2020s, warming to 29.60°C–30.70°C in the 2030s and 2040s.

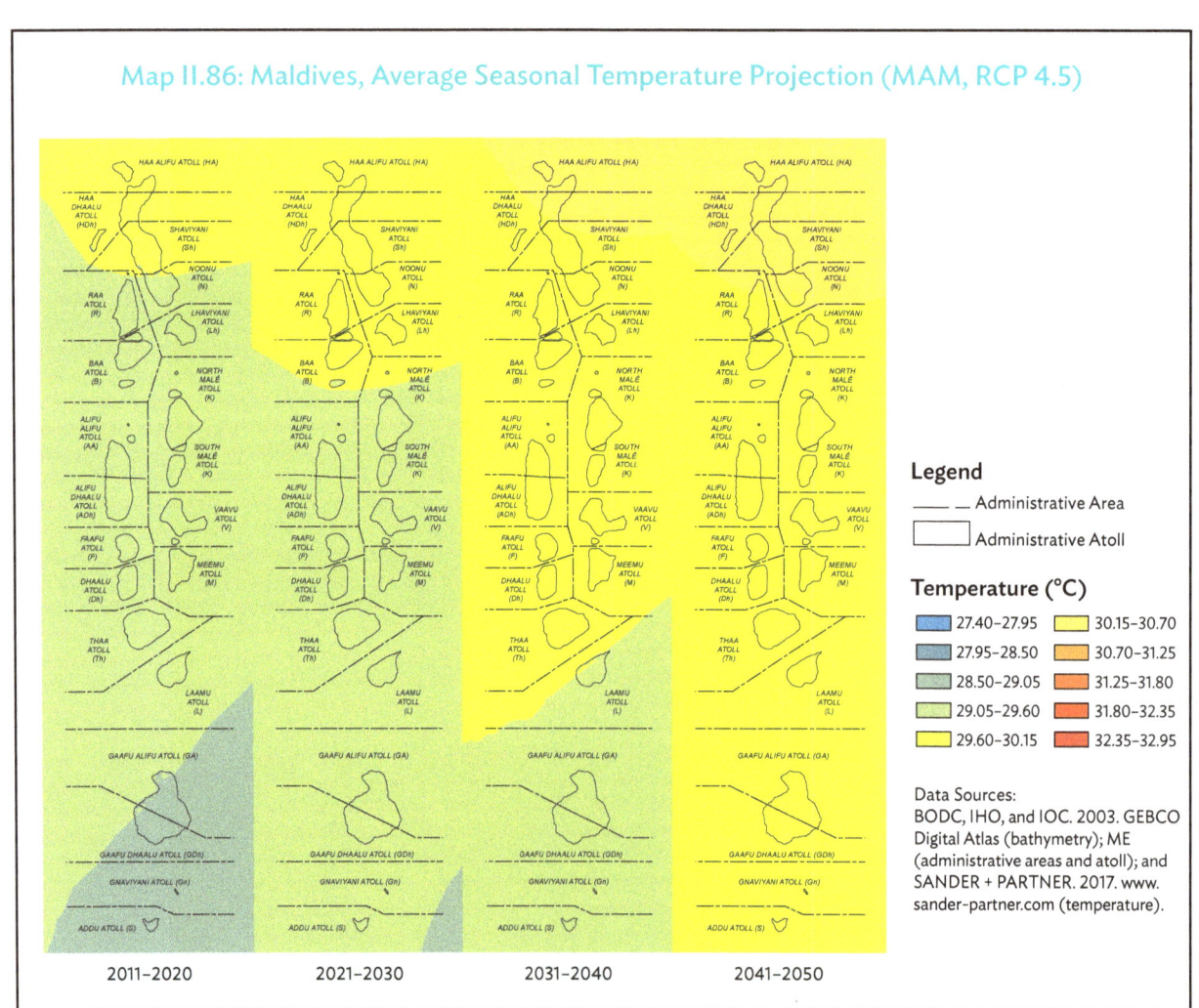

Map II.86: Maldives, Average Seasonal Temperature Projection (MAM, RCP 4.5)

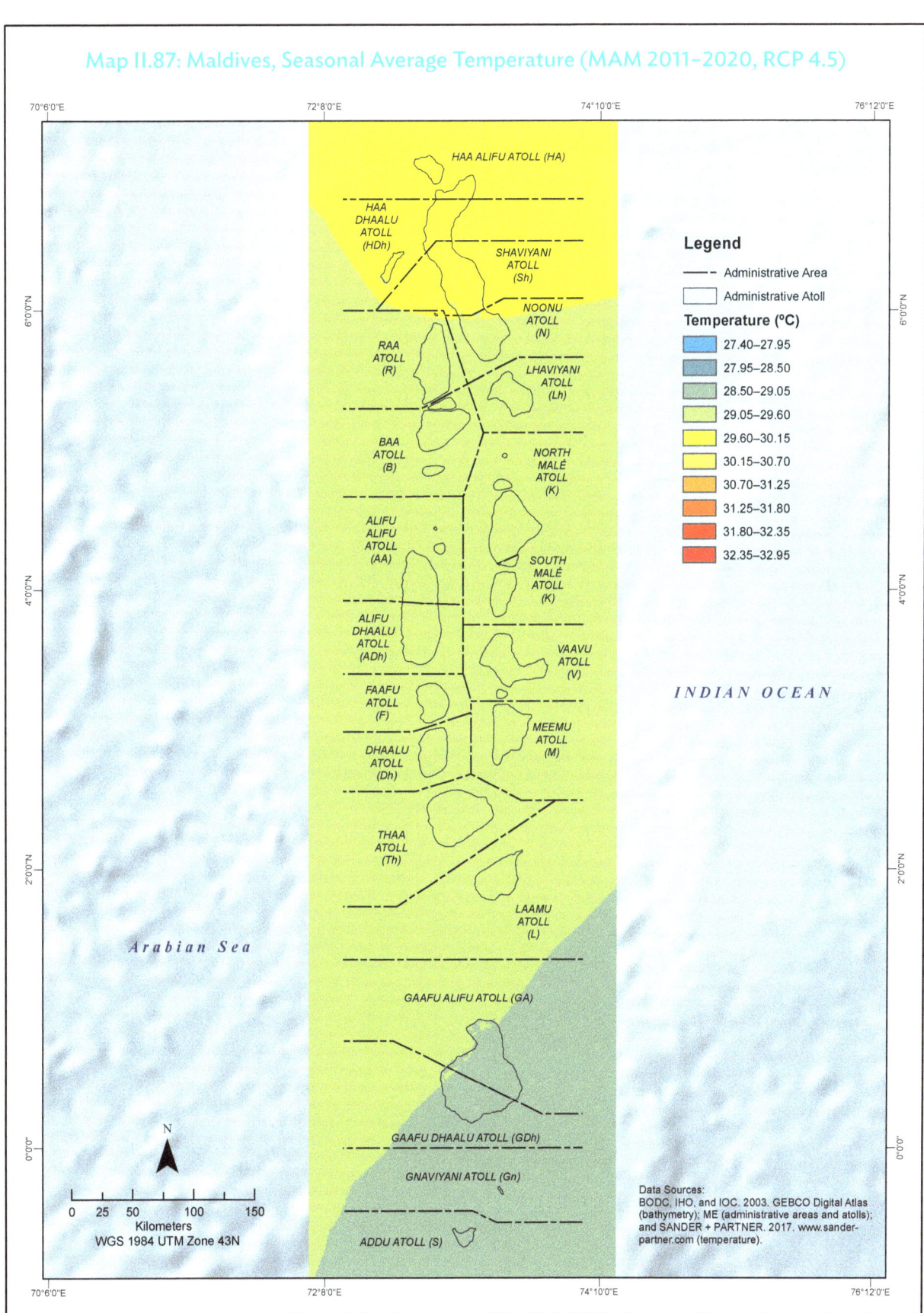

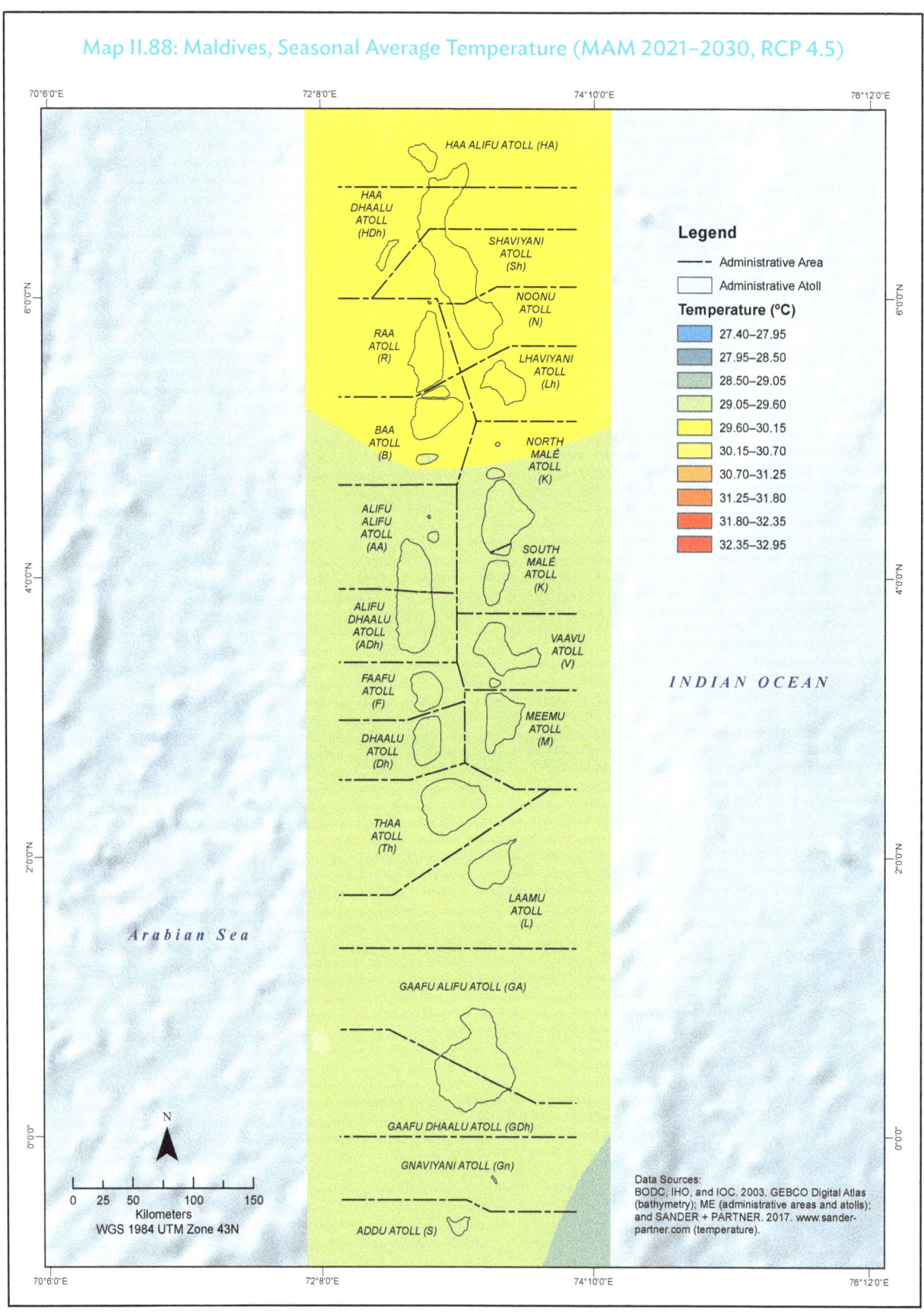

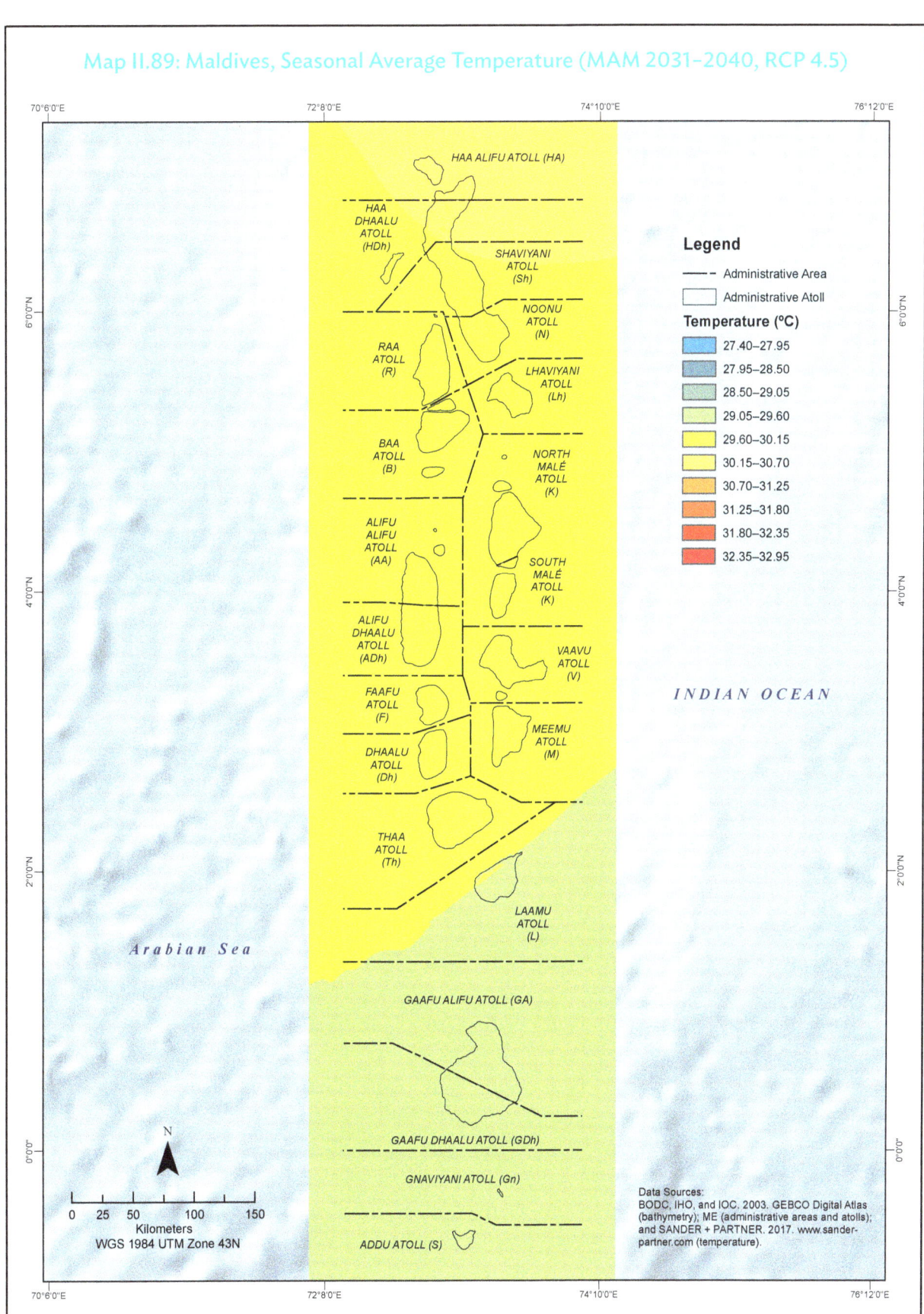

Map II.89: Maldives, Seasonal Average Temperature (MAM 2031–2040, RCP 4.5)

96 Multihazard Risk Atlas of Maldives—Climate and Geophysical Hazards

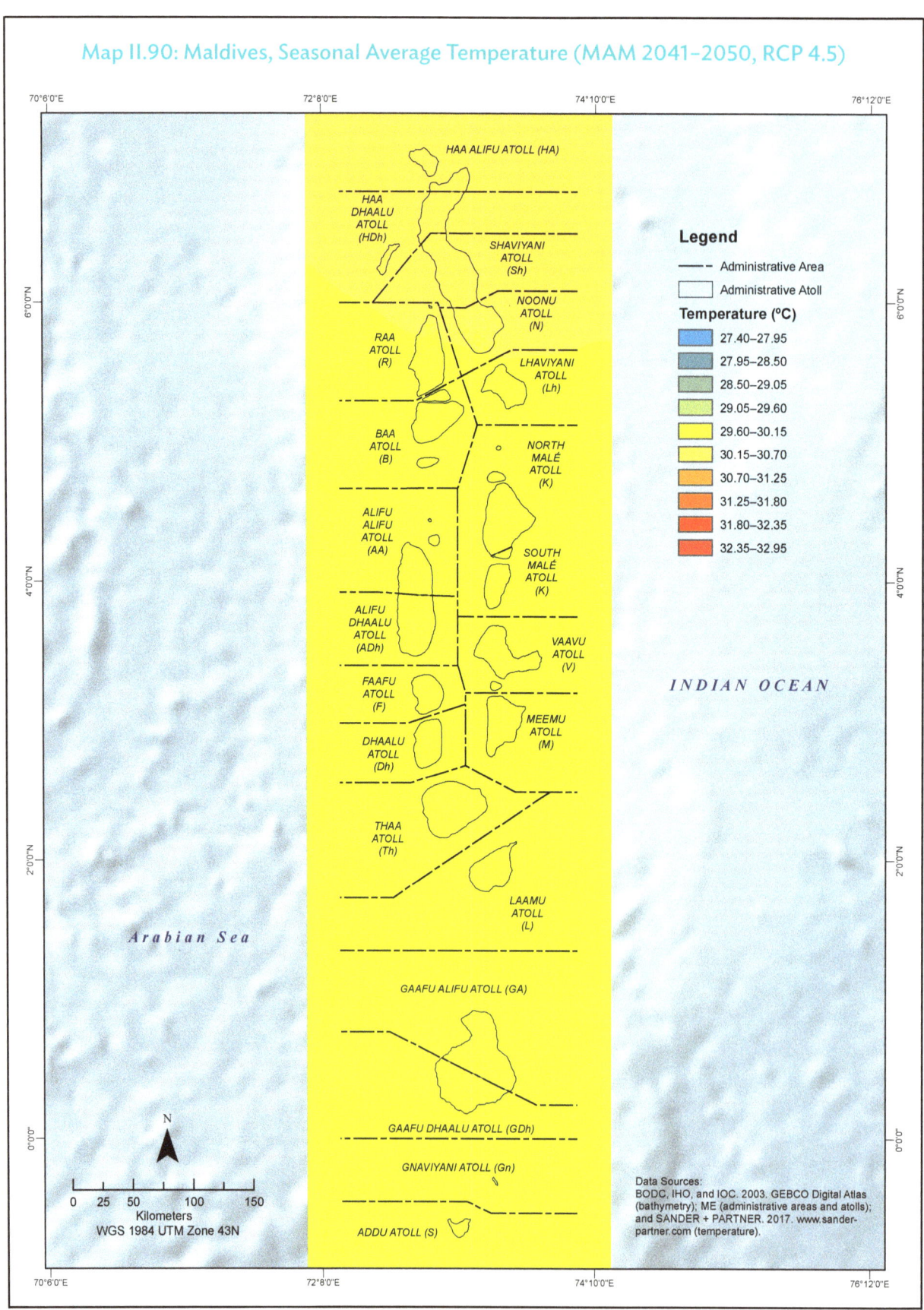

Map II.90: Maldives, Seasonal Average Temperature (MAM 2041–2050, RCP 4.5)

Average Seasonal Temperature Projection (JJA, RCP 4.5)

97

The projected average temperature during JJA shows warming in the future, with the temperature rising from 27.95°C–28.50°C to 29.05°C–29.60°C beginning in 2031, especially in the northern and middle atolls.

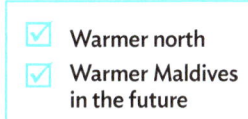

☑ Warmer north
☑ Warmer Maldives in the future

Map II.91: Maldives, Average Seasonal Temperature Projection (JJA, RCP 4.5)

Legend
— Administrative Area
☐ Administrative Atoll

Temperature (°C)
- 27.40–27.95
- 27.95–28.50
- 28.50–29.05
- 29.05–29.60
- 29.60–30.15
- 30.15–30.70
- 30.70–31.25
- 31.25–31.80
- 31.80–32.35
- 32.35–32.95

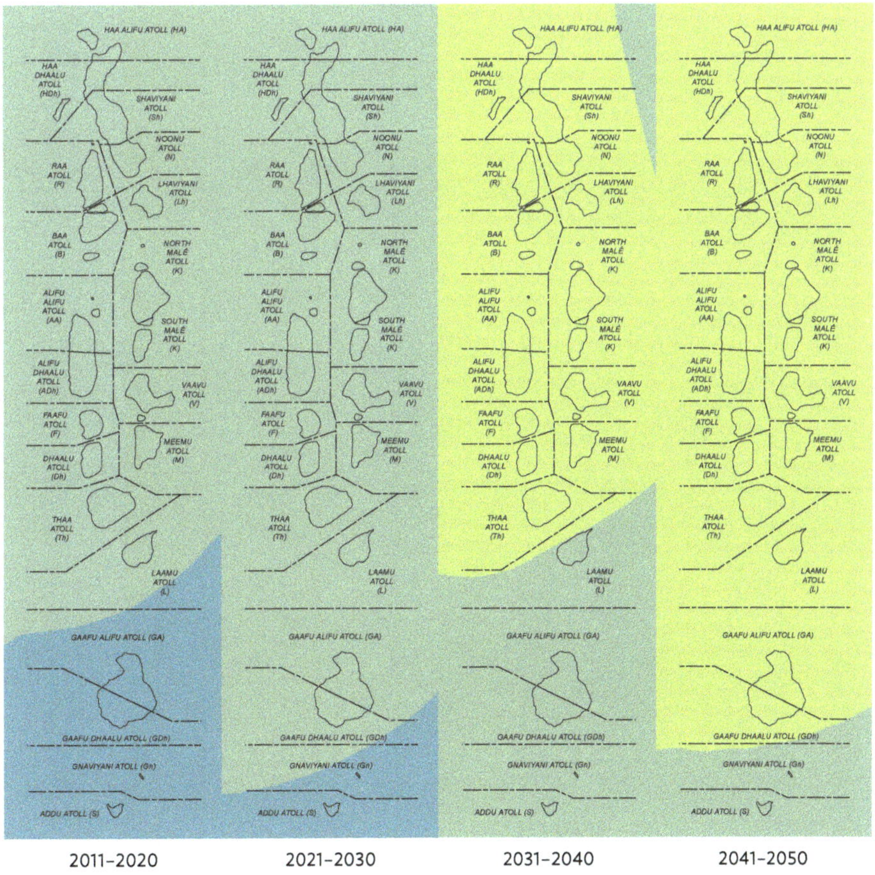

Data Sources:
BODC, IHO, and IOC. 2003. GEBCO Digital Atlas (bathymetry); ME (administrative areas and atoll); and SANDER + PARTNER. 2017. www.sander-partner.com (temperature).

2011–2020 | 2021–2030 | 2031–2040 | 2041–2050

98 Multihazard Risk Atlas of Maldives—Climate and Geophysical Hazards

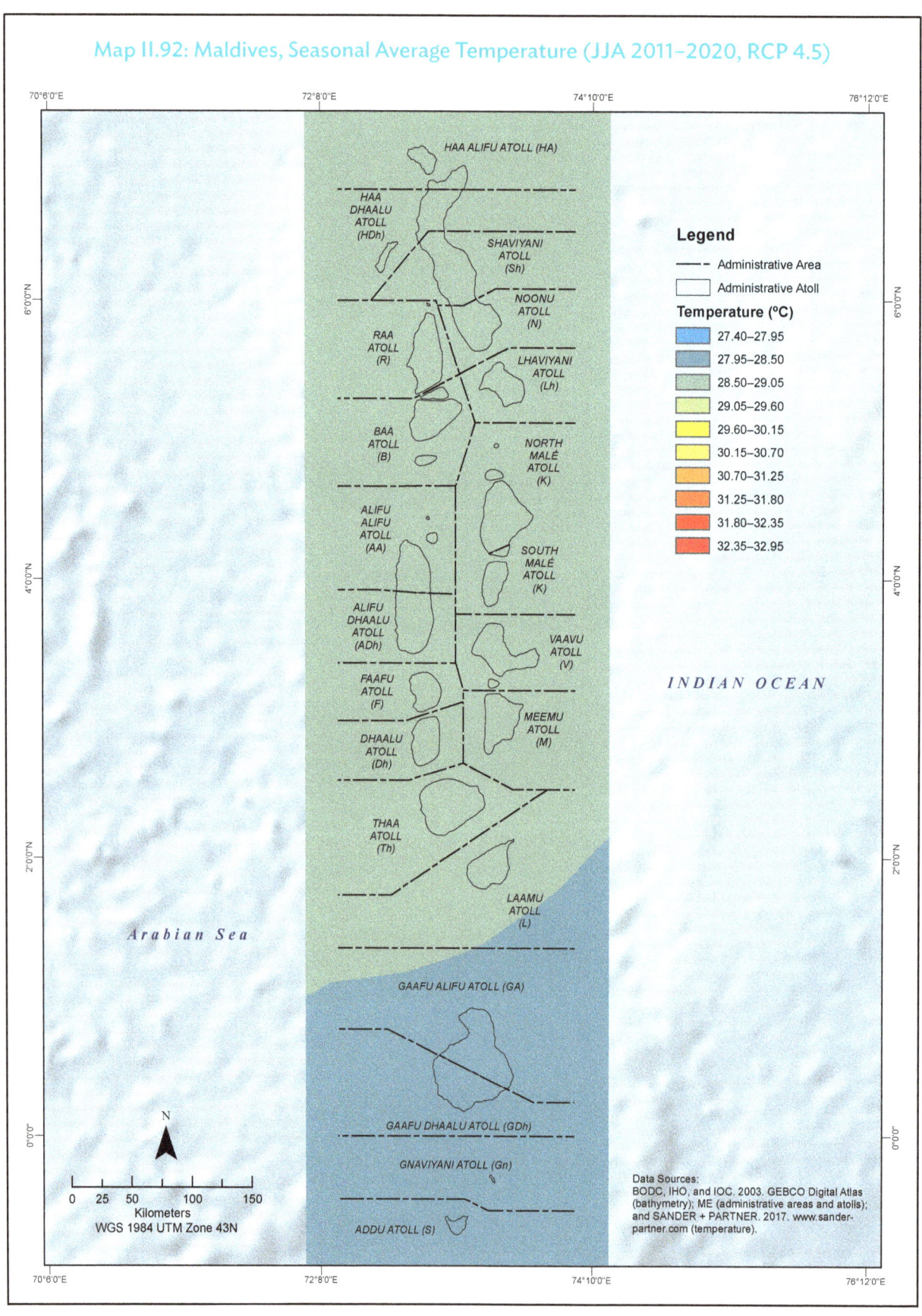

Future Climate

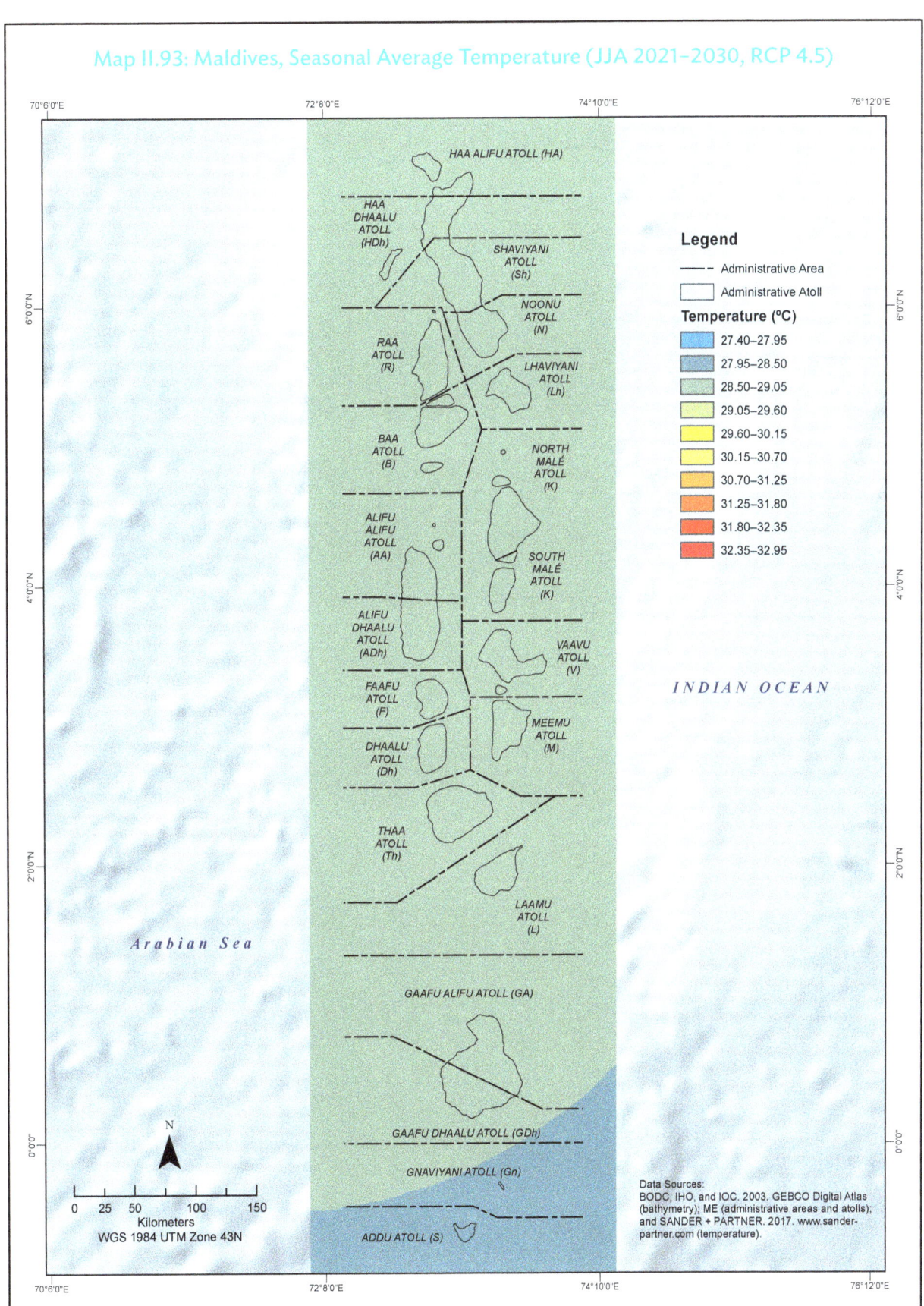

Map II.93: Maldives, Seasonal Average Temperature (JJA 2021–2030, RCP 4.5)

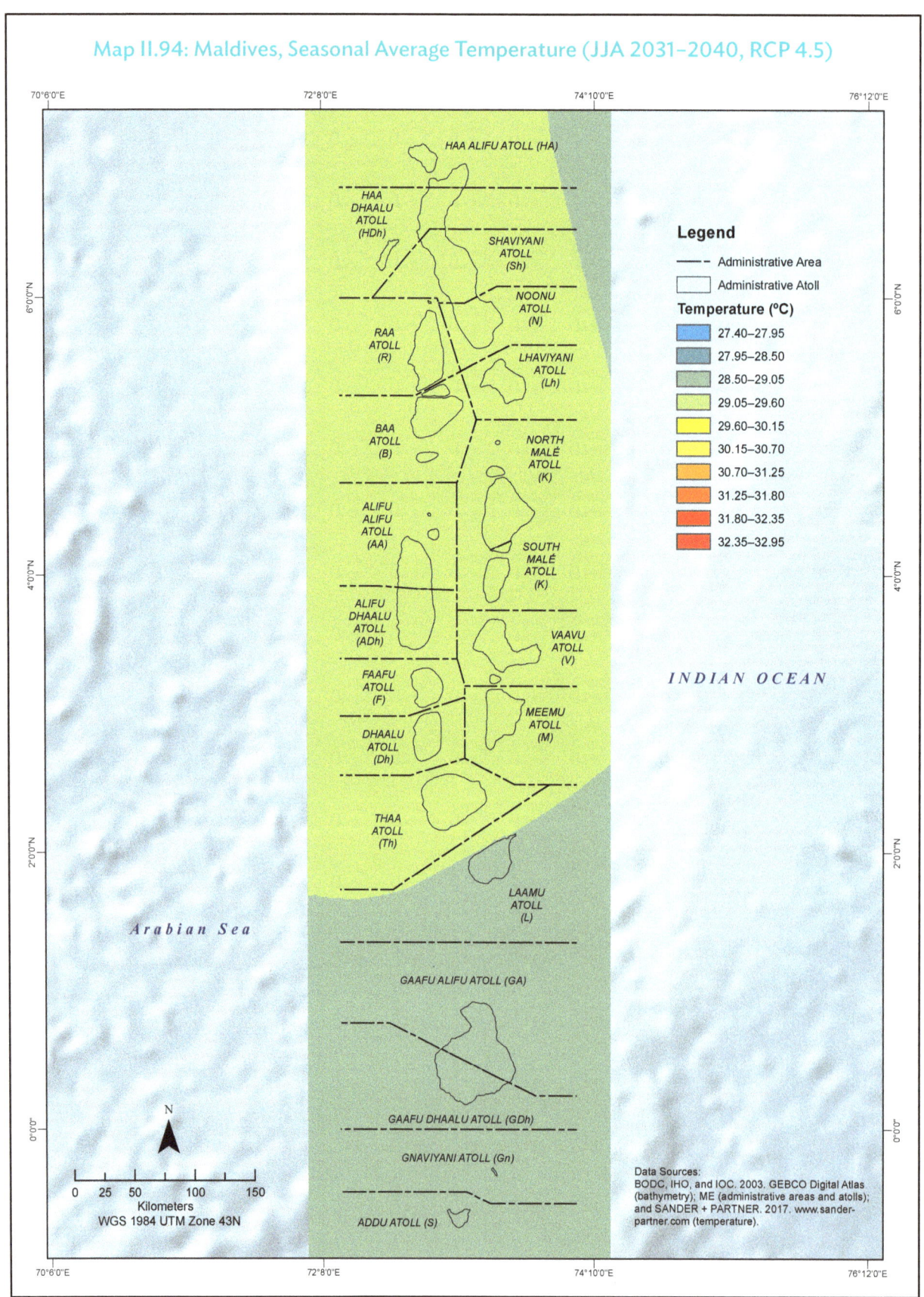

Map II.94: Maldives, Seasonal Average Temperature (JJA 2031–2040, RCP 4.5)

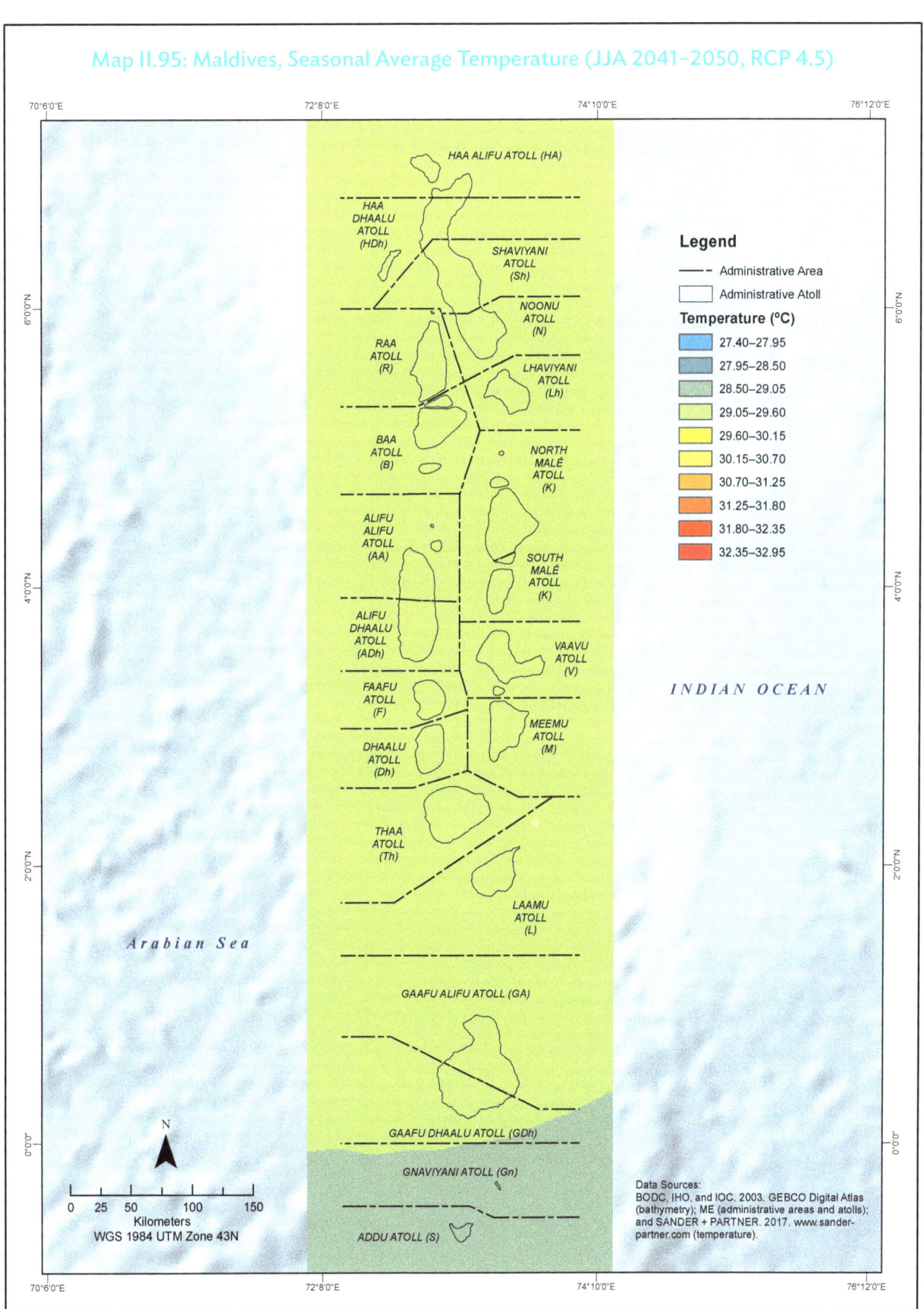

Future Climate 101

Map II.95: Maldives, Seasonal Average Temperature (JJA 2041–2050, RCP 4.5)

Average Seasonal Temperature Projection (SON, RCP 4.5)

- ☑ Warmer north
- ☑ Warmer Maldives in the future
- ☑ SON are the coldest months

SON are historically the coldest months of the year. The projected average temperature for SON shows they will continue to be Maldives' coldest months. However, there is a general increase in the average temperature from 27.95°C–28.50°C in 2011–2020 to 28.50°C–29.05°C in 2041–2050.

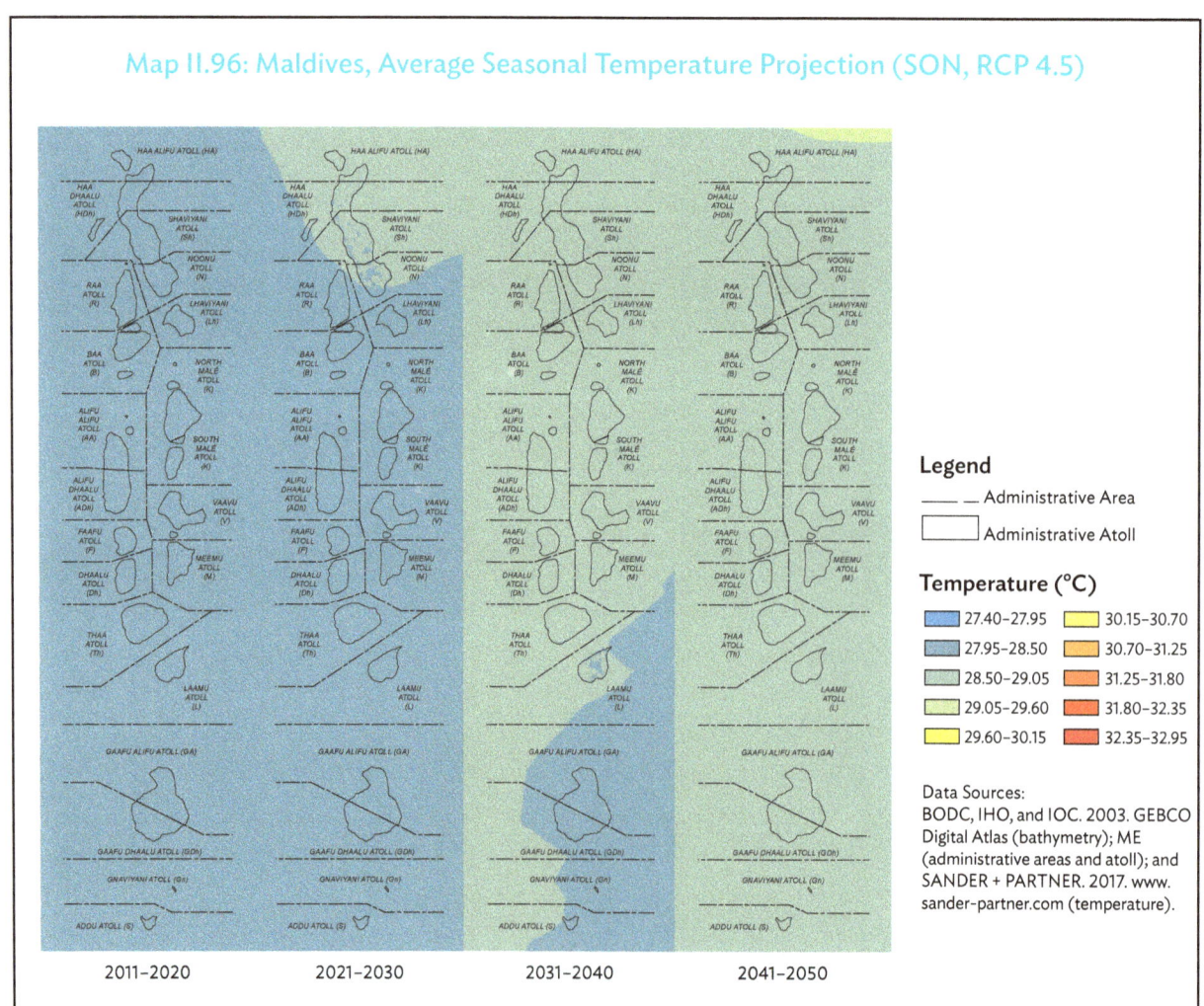

Map II.96: Maldives, Average Seasonal Temperature Projection (SON, RCP 4.5)

Future Climate 103

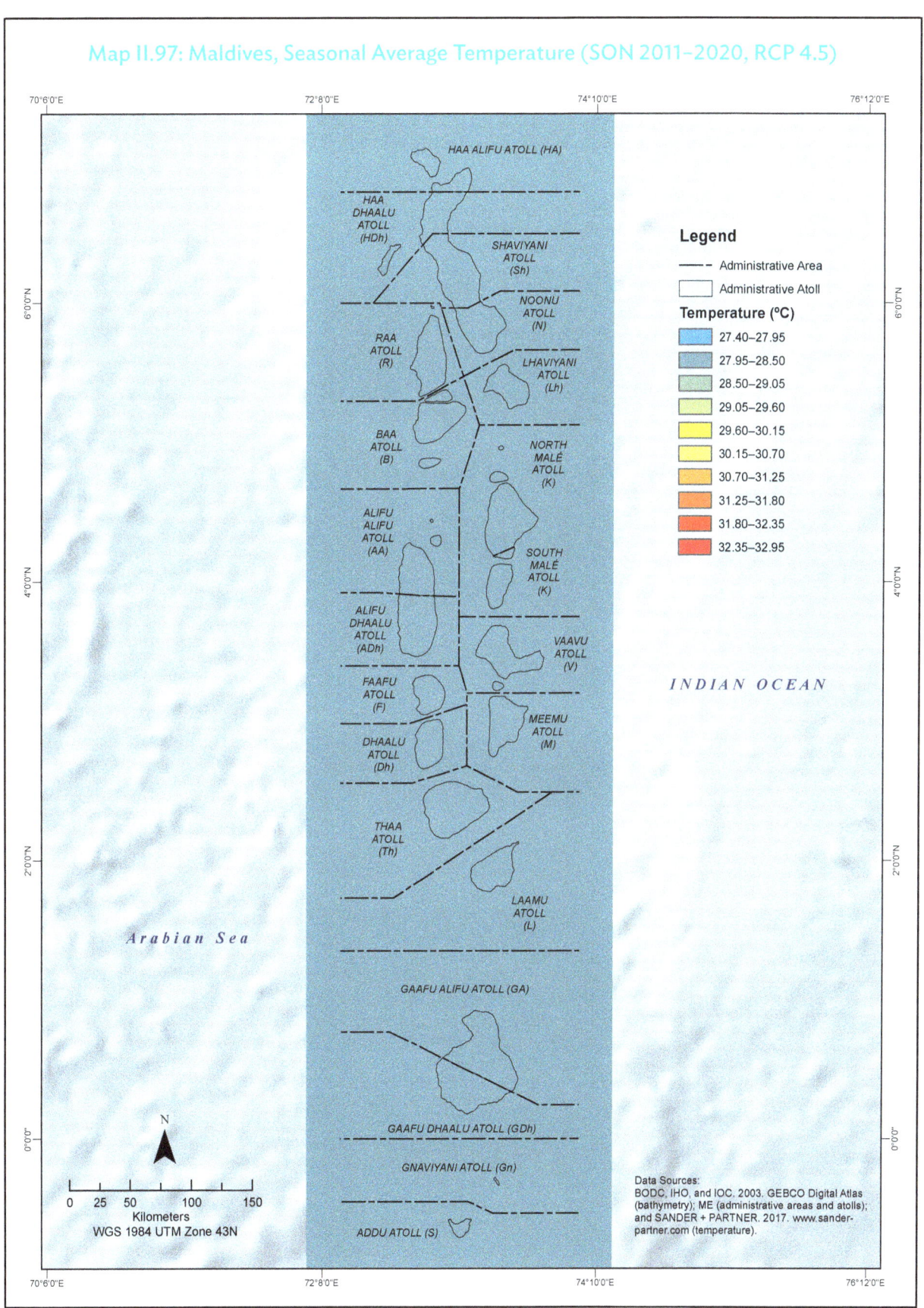

Map II.97: Maldives, Seasonal Average Temperature (SON 2011–2020, RCP 4.5)

104 Multihazard Risk Atlas of Maldives—Climate and Geophysical Hazards

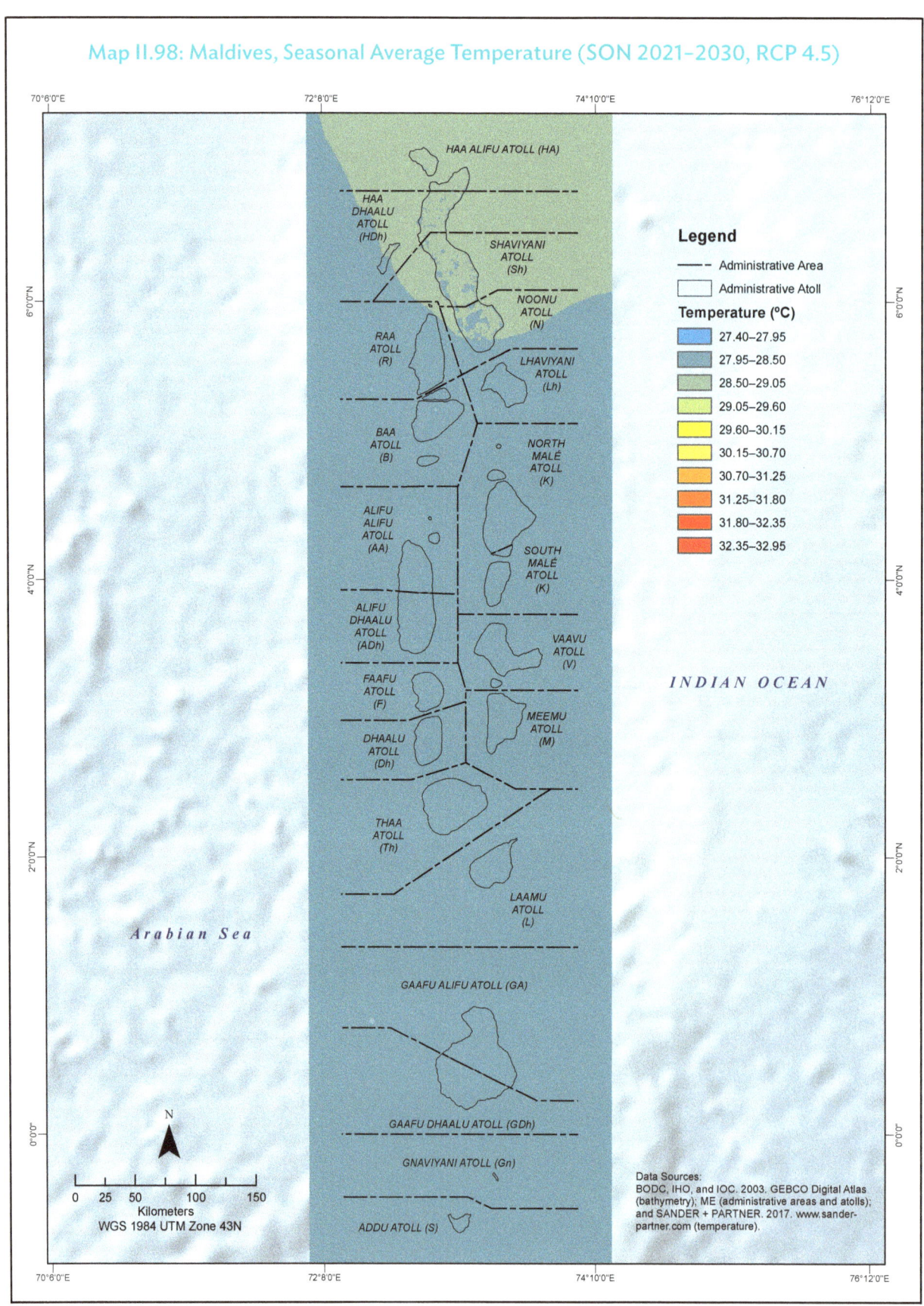

Map II.98: Maldives, Seasonal Average Temperature (SON 2021–2030, RCP 4.5)

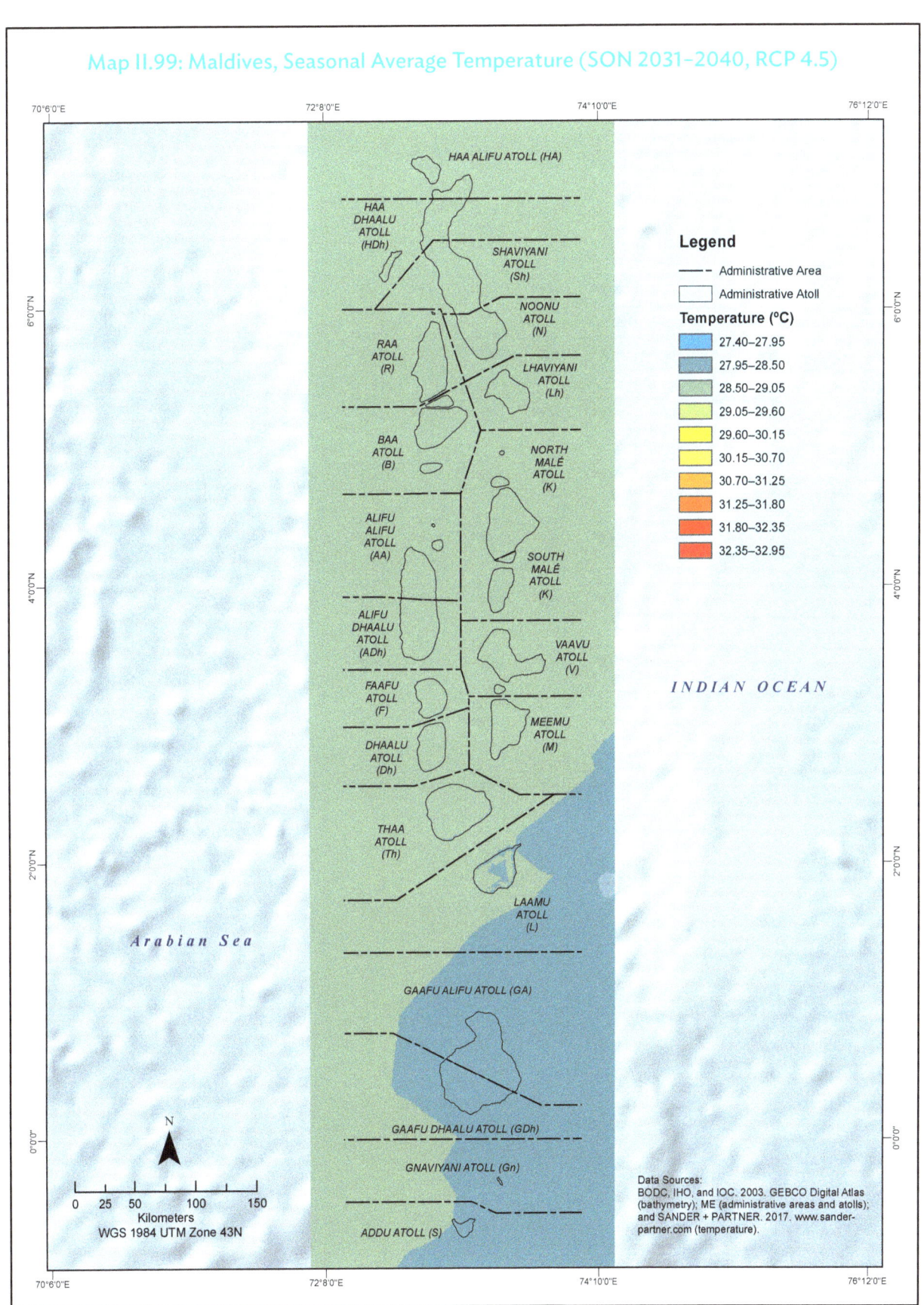

Map II.99: Maldives, Seasonal Average Temperature (SON 2031–2040, RCP 4.5)

106 Multihazard Risk Atlas of Maldives—Climate and Geophysical Hazards

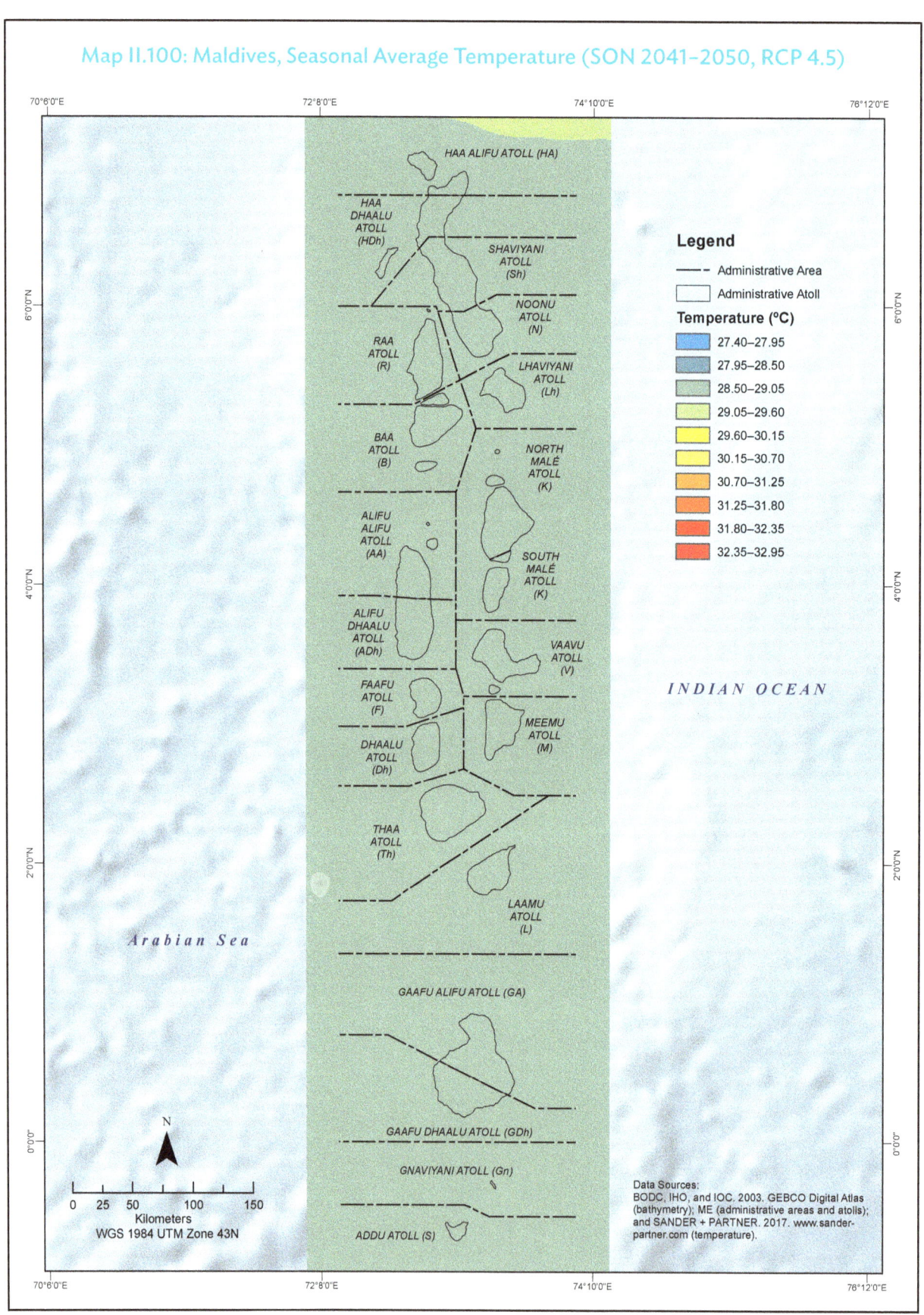

Map II.100: Maldives, Seasonal Average Temperature (SON 2041–2050, RCP 4.5)

Average Annual Temperature Projection (RCP 8.5)

107

In the RCP 8.5 scenario, Maldives is projected to be warmer in the 2030s and 2040s, especially in the northern atolls. Average temperature will increase from 29.05°C–29.60°C in the 2020s and 2030s to 29.60°C–30.15°C in the 2040s.

☑ Warmer north
☑ Warmer Maldives in the future

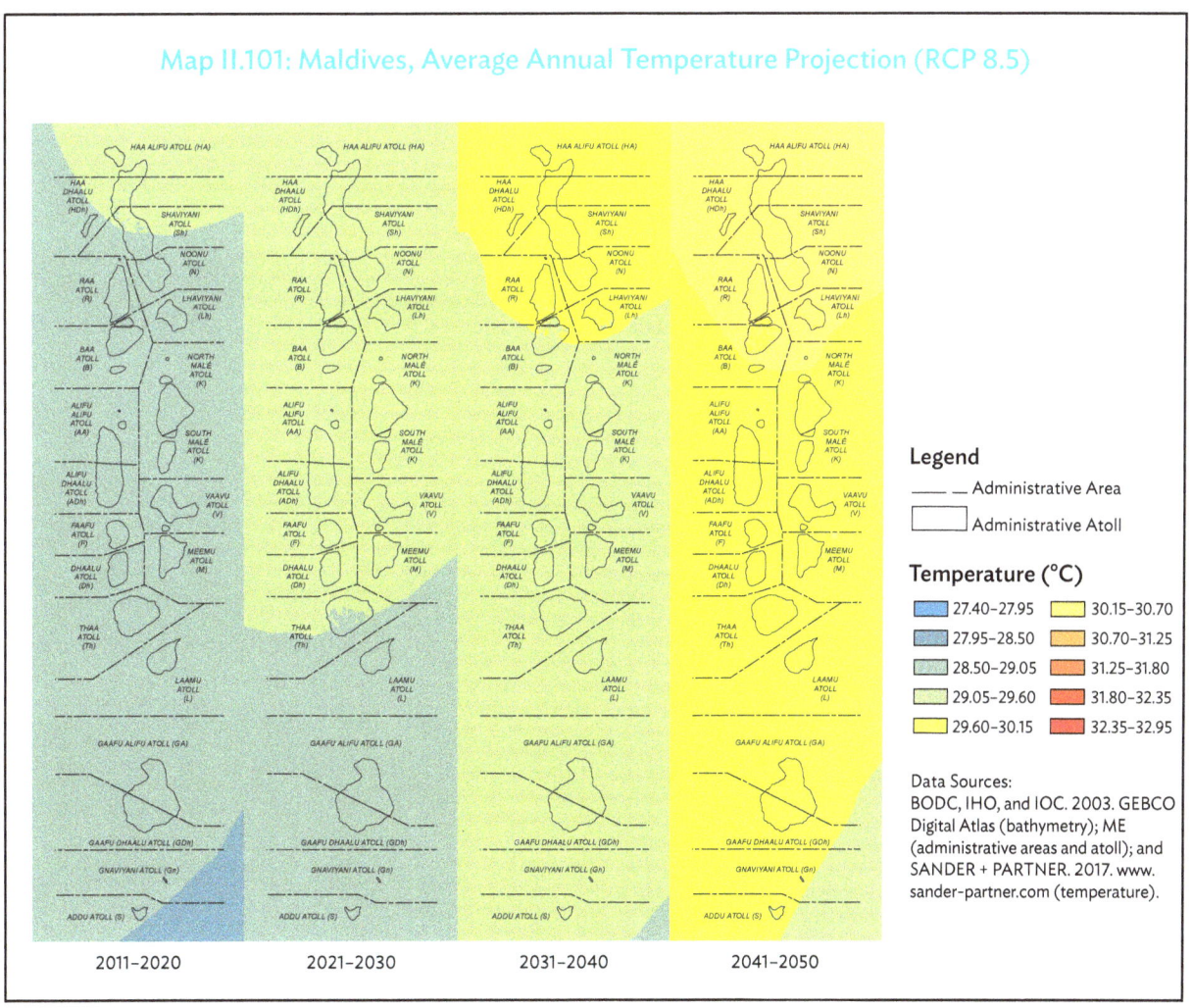

Map II.101: Maldives, Average Annual Temperature Projection (RCP 8.5)

Legend
— Administrative Area
☐ Administrative Atoll

Temperature (°C)
- 27.40–27.95
- 27.95–28.50
- 28.50–29.05
- 29.05–29.60
- 29.60–30.15
- 30.15–30.70
- 30.70–31.25
- 31.25–31.80
- 31.80–32.35
- 32.35–32.95

Data Sources:
BODC, IHO, and IOC. 2003. GEBCO Digital Atlas (bathymetry); ME (administrative areas and atoll); and SANDER + PARTNER. 2017. www.sander-partner.com (temperature).

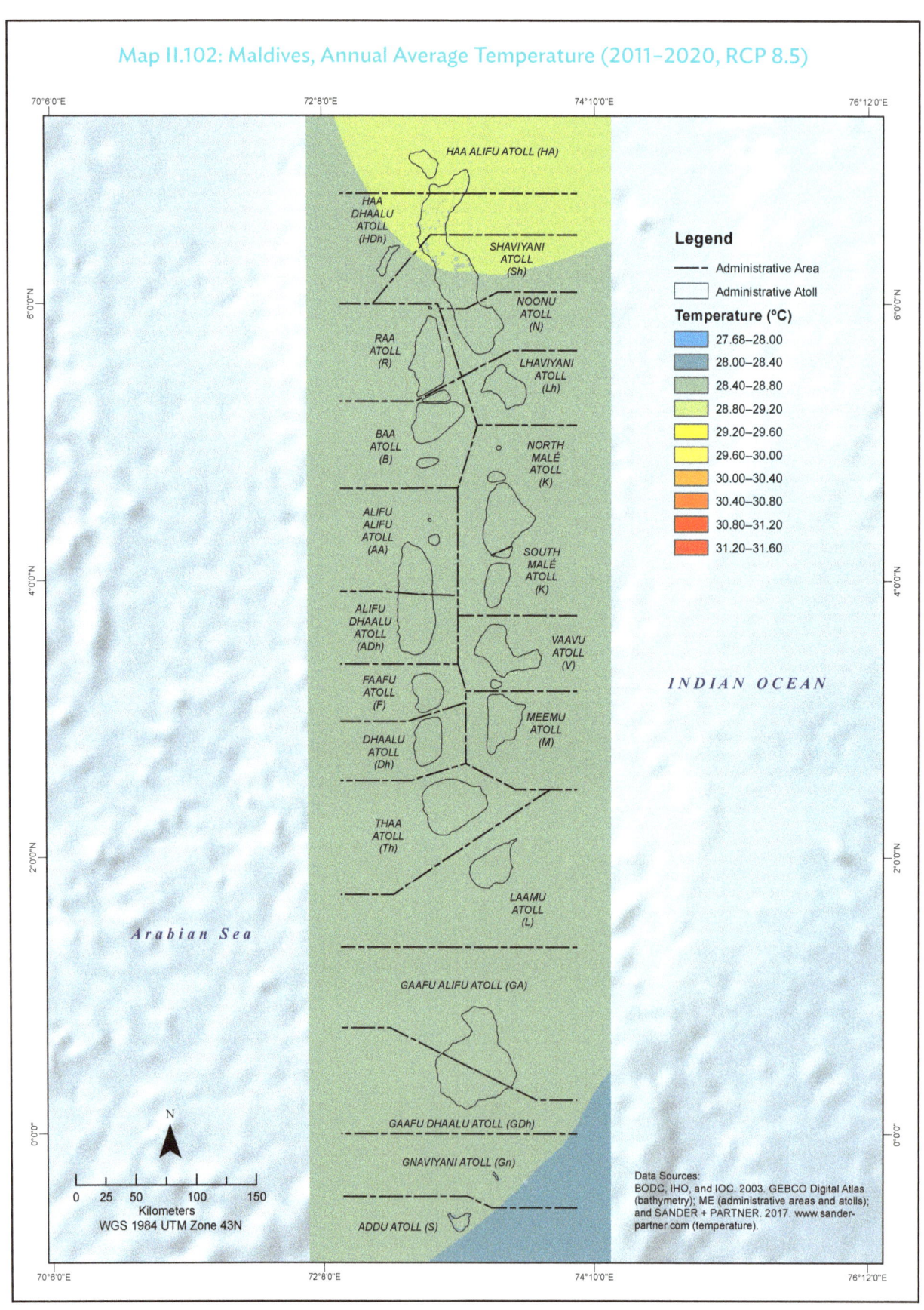

Map II.102: Maldives, Annual Average Temperature (2011–2020, RCP 8.5)

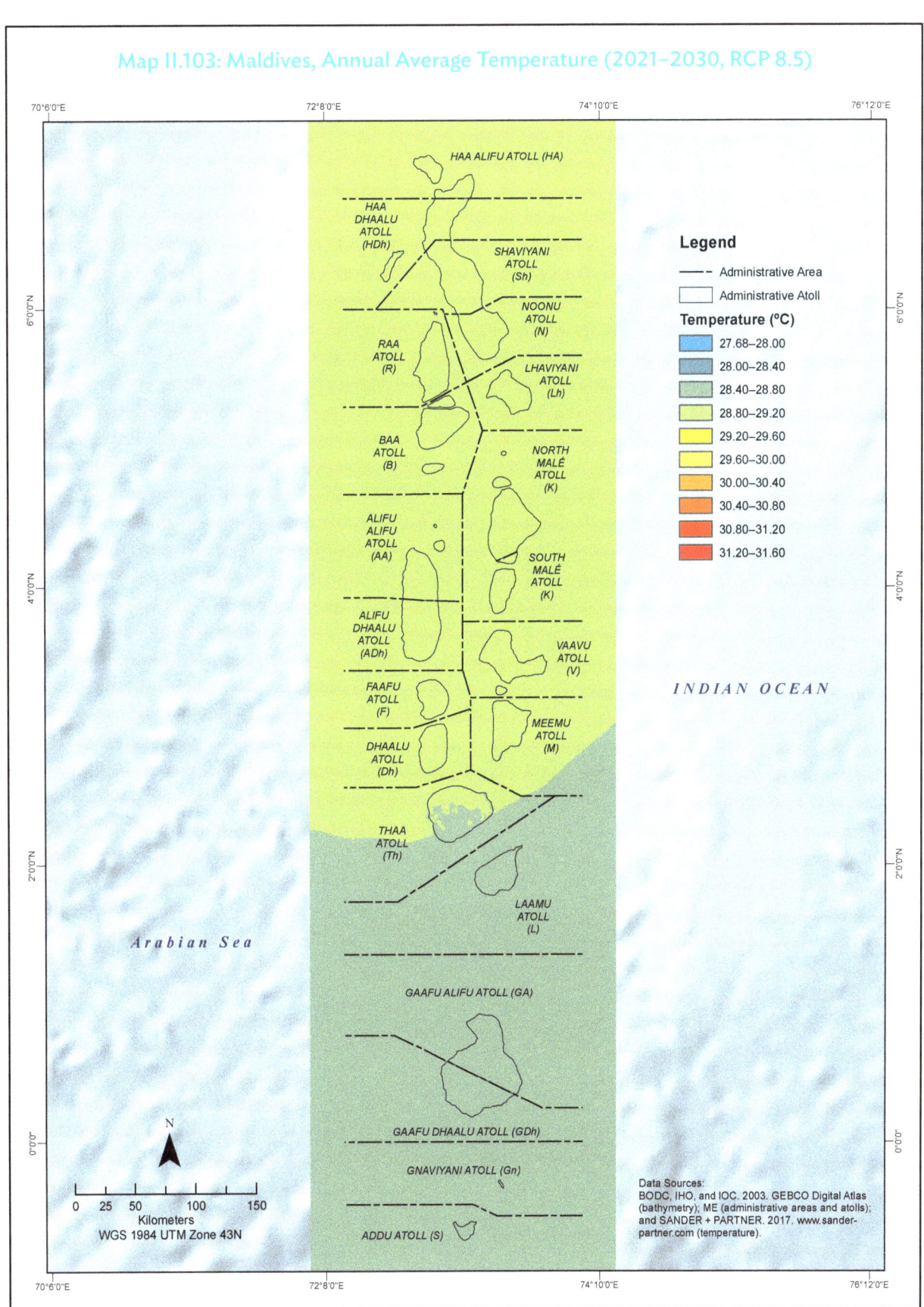

Map II.103: Maldives, Annual Average Temperature (2021–2030, RCP 8.5)

110 Multihazard Risk Atlas of Maldives—Climate and Geophysical Hazards

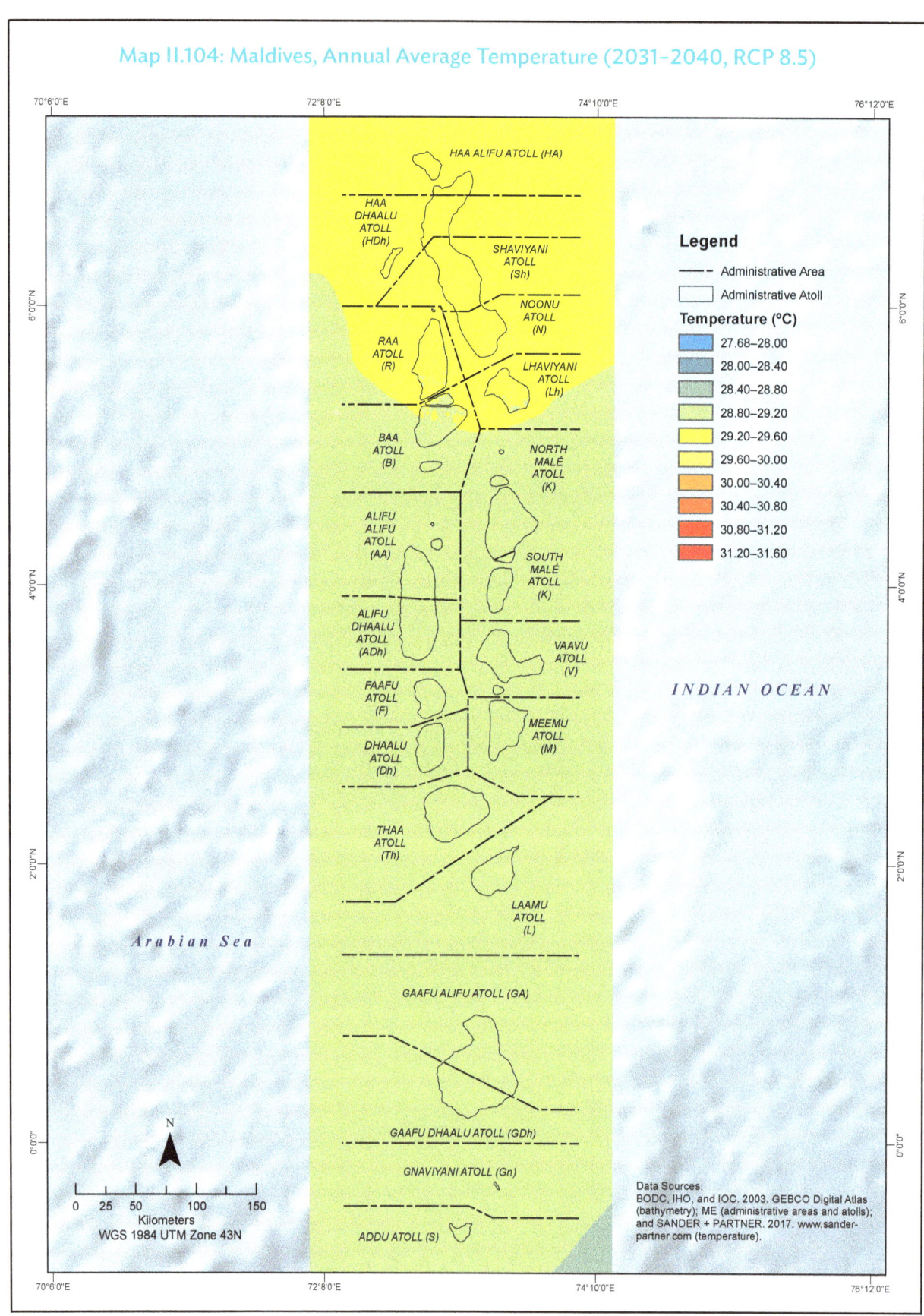

Map II.104: Maldives, Annual Average Temperature (2031–2040, RCP 8.5)

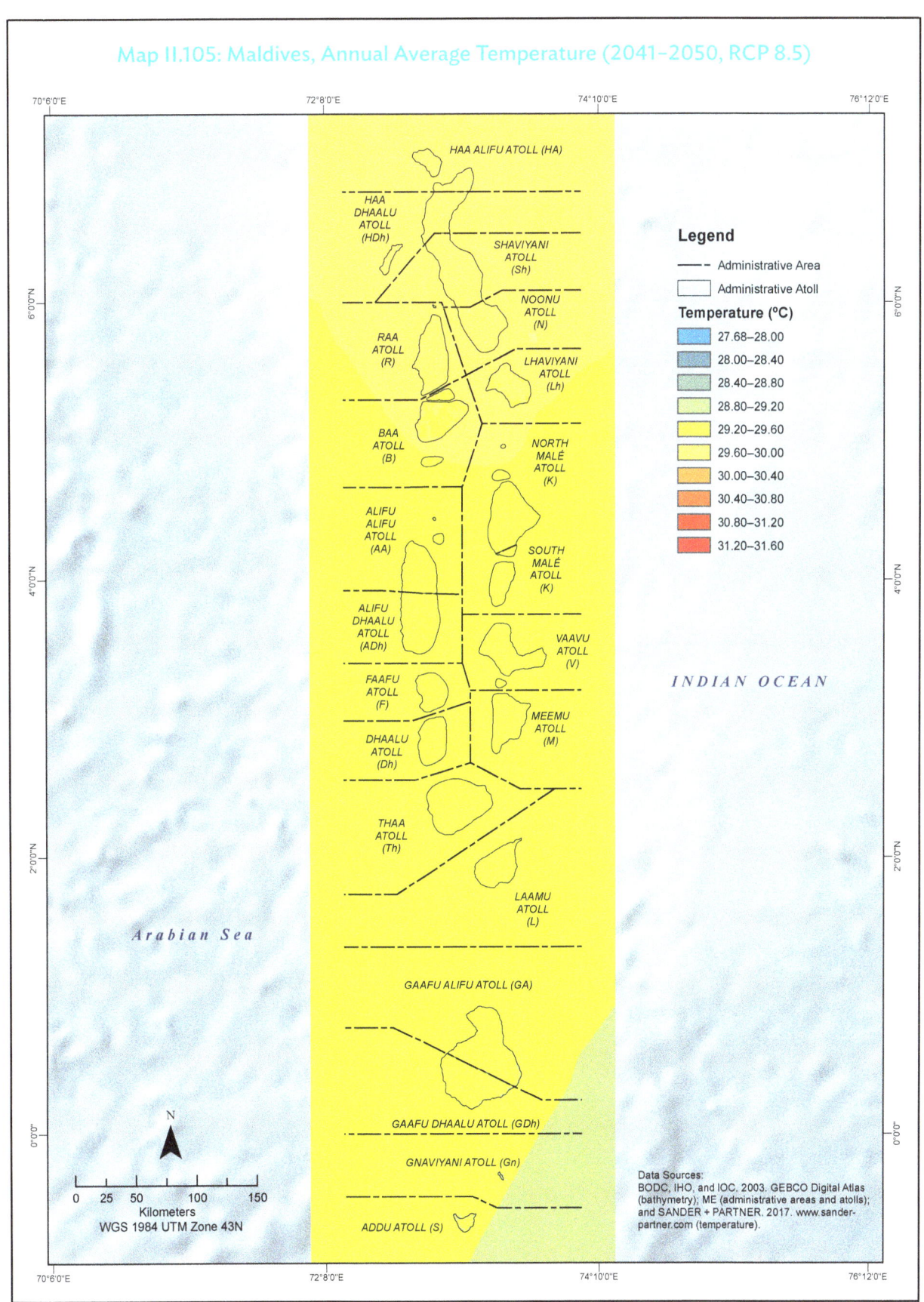

Average Seasonal Temperature Projection (DJF, RCP 8.5)

- ✓ Warmer north
- ✓ Warming in northern Maldives in the future
- ✓ RCP 8.5 projected higher temperature than RCP 4.5

The RCP 8.5 average temperature projection for DJF (average range of 28.50°C–29.05°C) shows no spatial variation in the temperature across Maldives from 2011 to 2030. The temperature will start to rise in the 2030s, reaching 29.05°C–29.60°C. By the 2040s, the northern and middle portion of Maldives will be warmer, with an average range of 29.60°C–30.15°C.

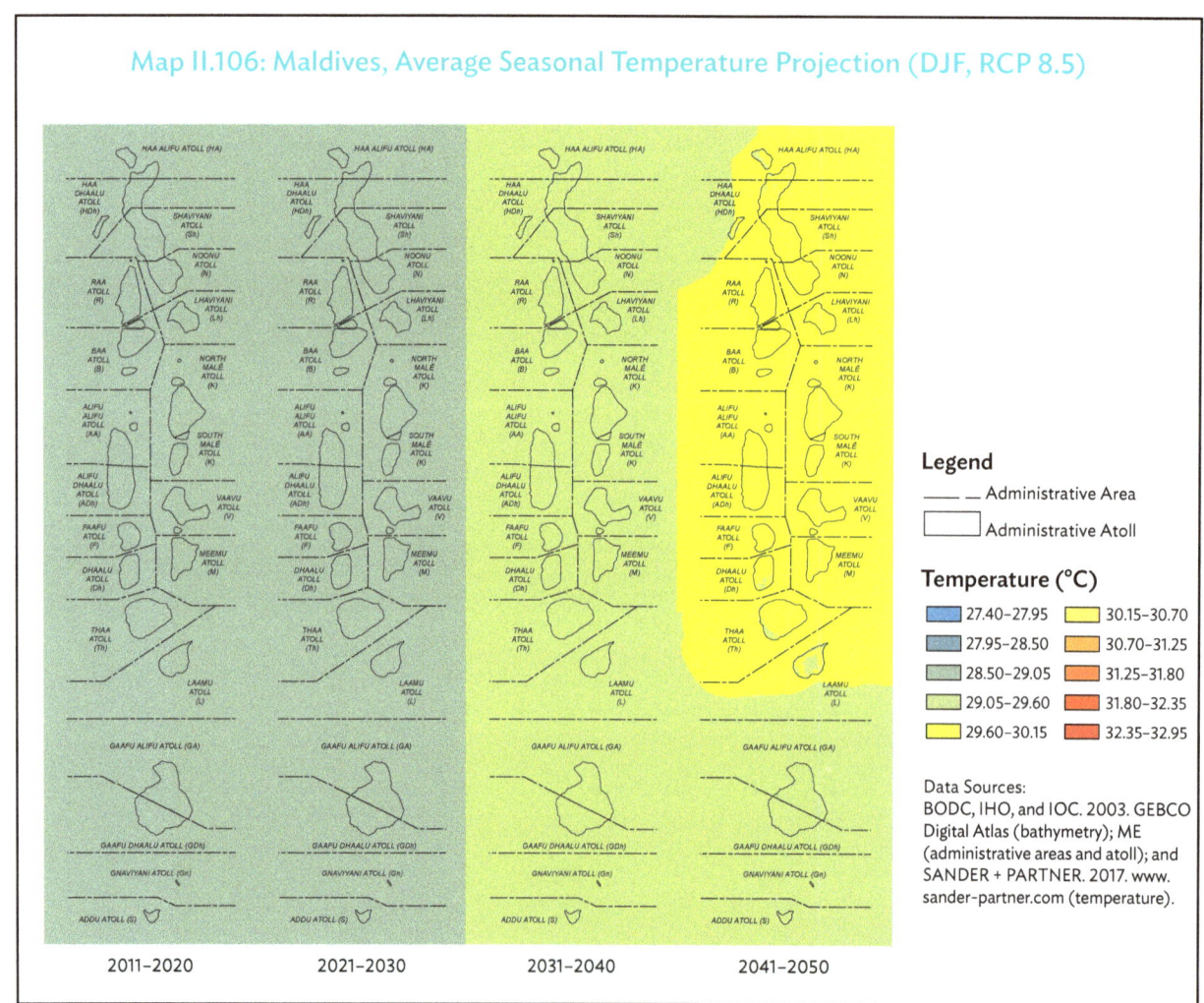

Map II.106: Maldives, Average Seasonal Temperature Projection (DJF, RCP 8.5)

Data Sources: BODC, IHO, and IOC. 2003. GEBCO Digital Atlas (bathymetry); ME (administrative areas and atoll); and SANDER + PARTNER. 2017. www.sander-partner.com (temperature).

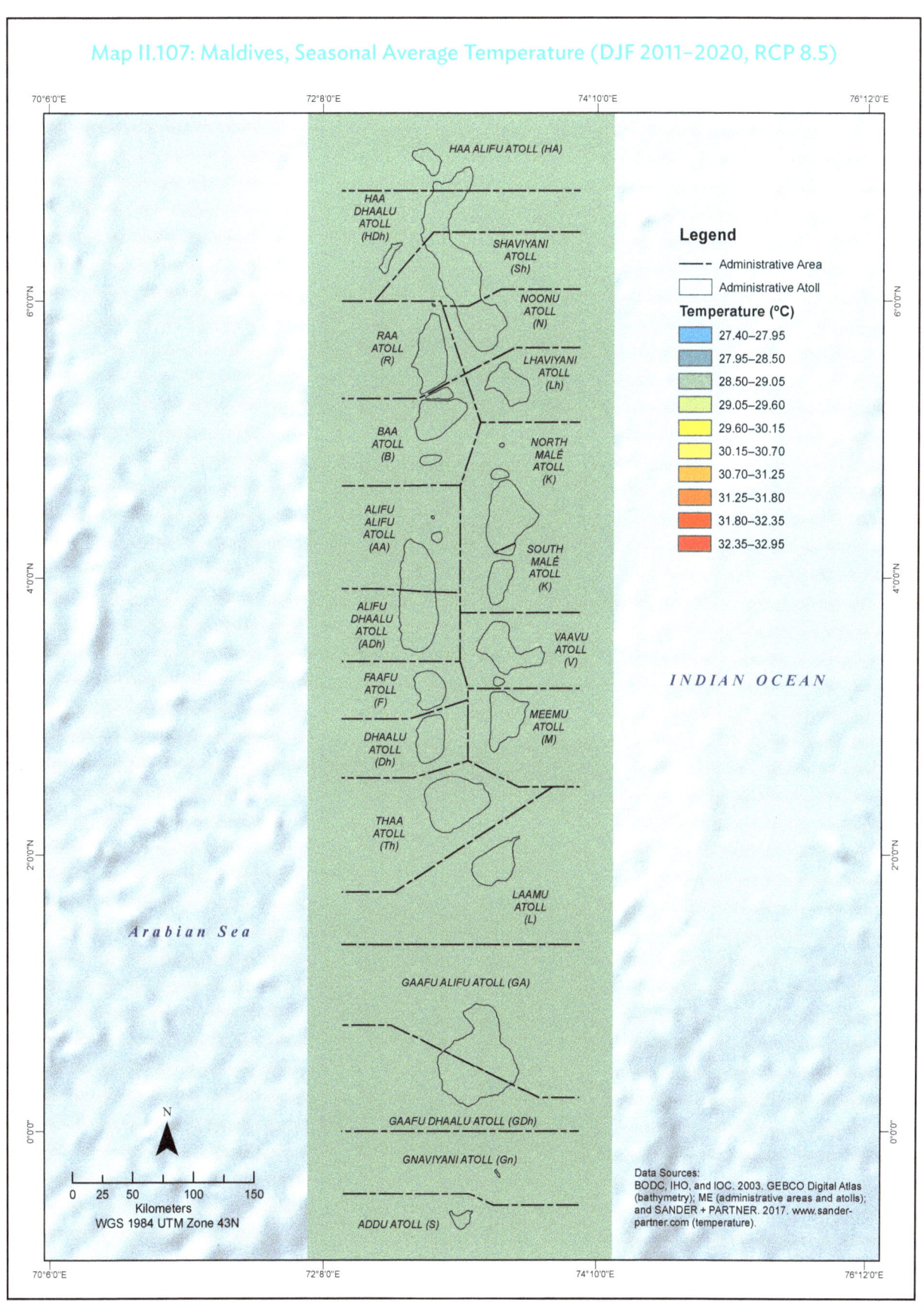

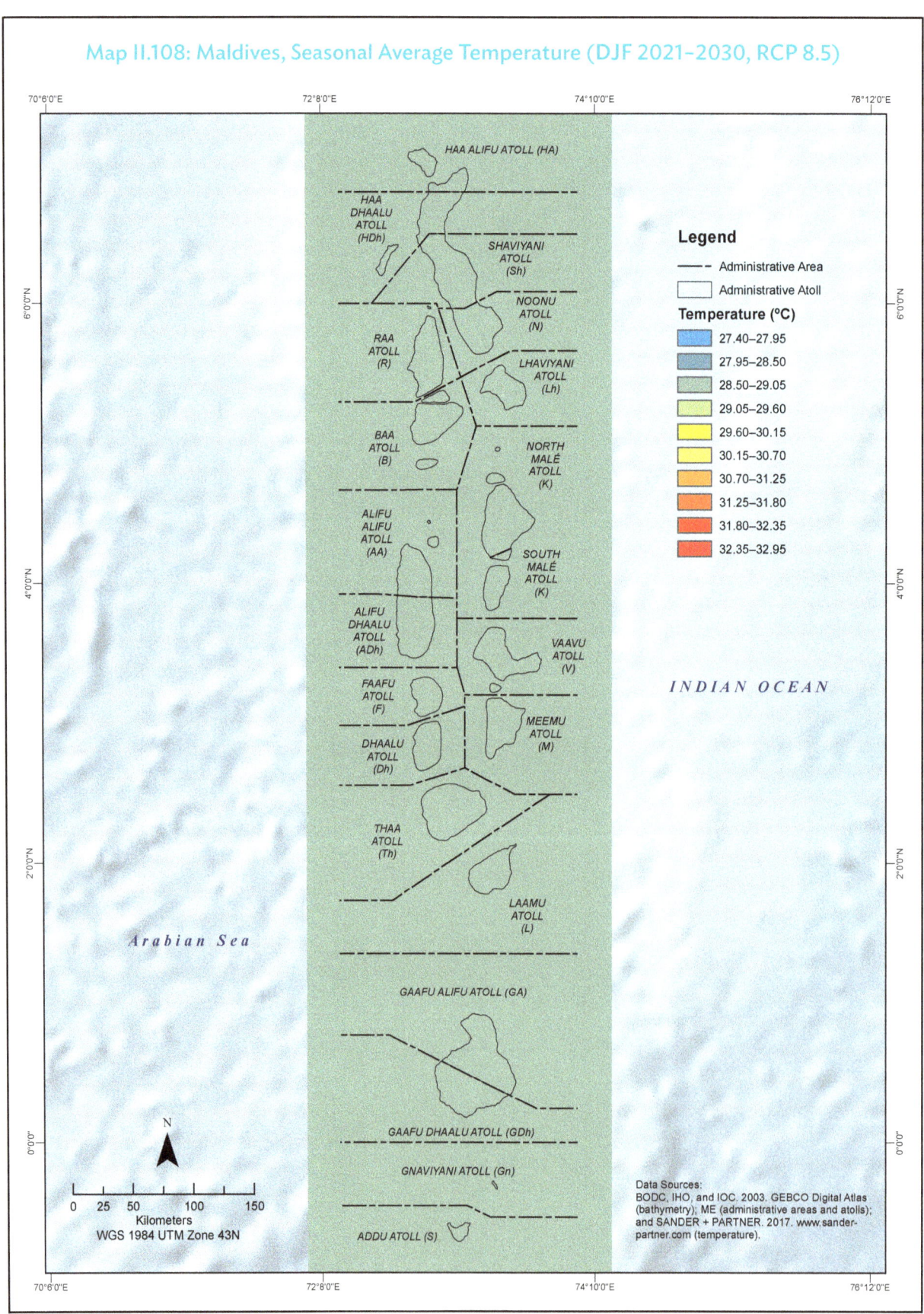

Map II.108: Maldives, Seasonal Average Temperature (DJF 2021–2030, RCP 8.5)

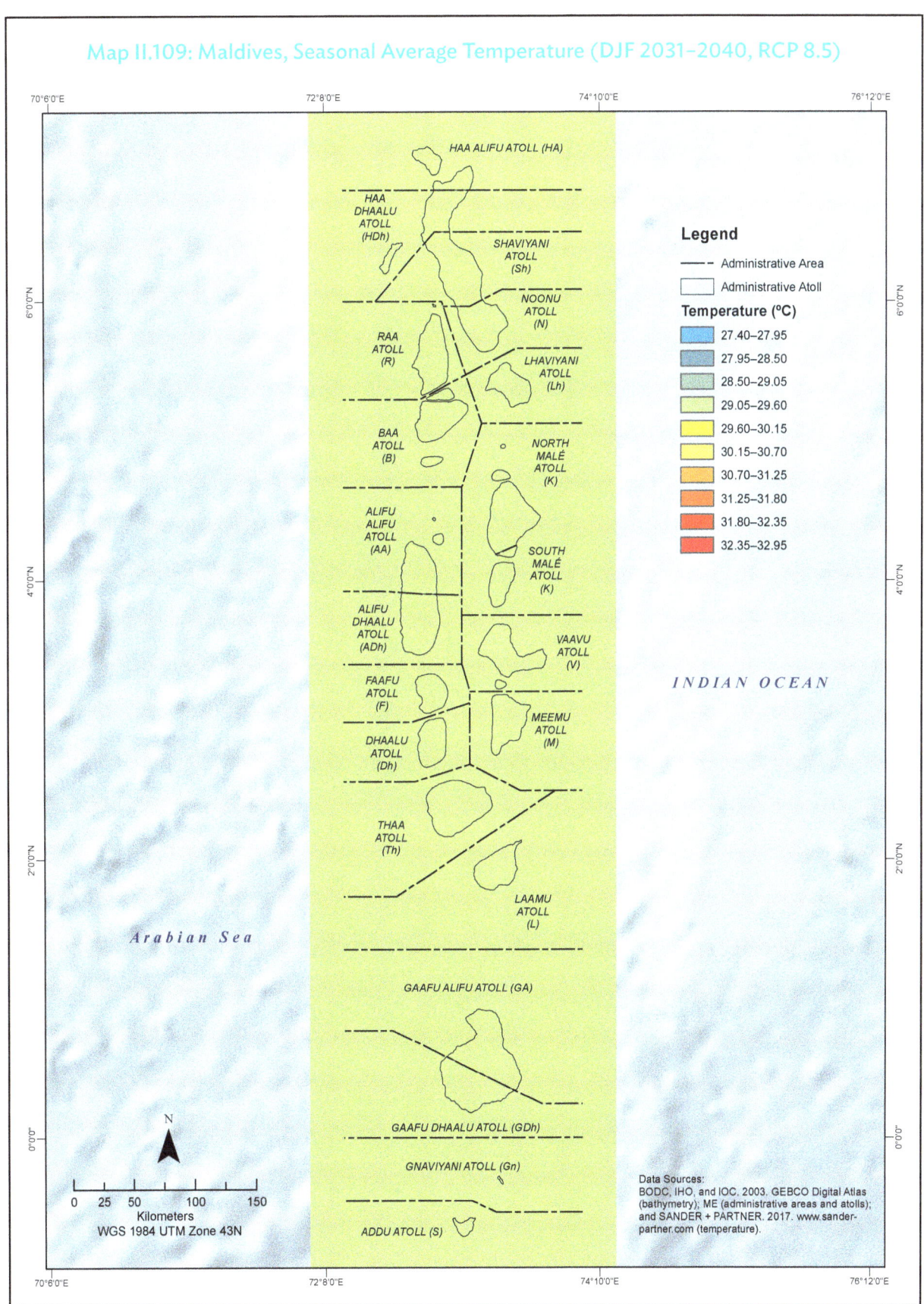

Map II.109: Maldives, Seasonal Average Temperature (DJF 2031–2040, RCP 8.5)

116 Multihazard Risk Atlas of Maldives—Climate and Geophysical Hazards

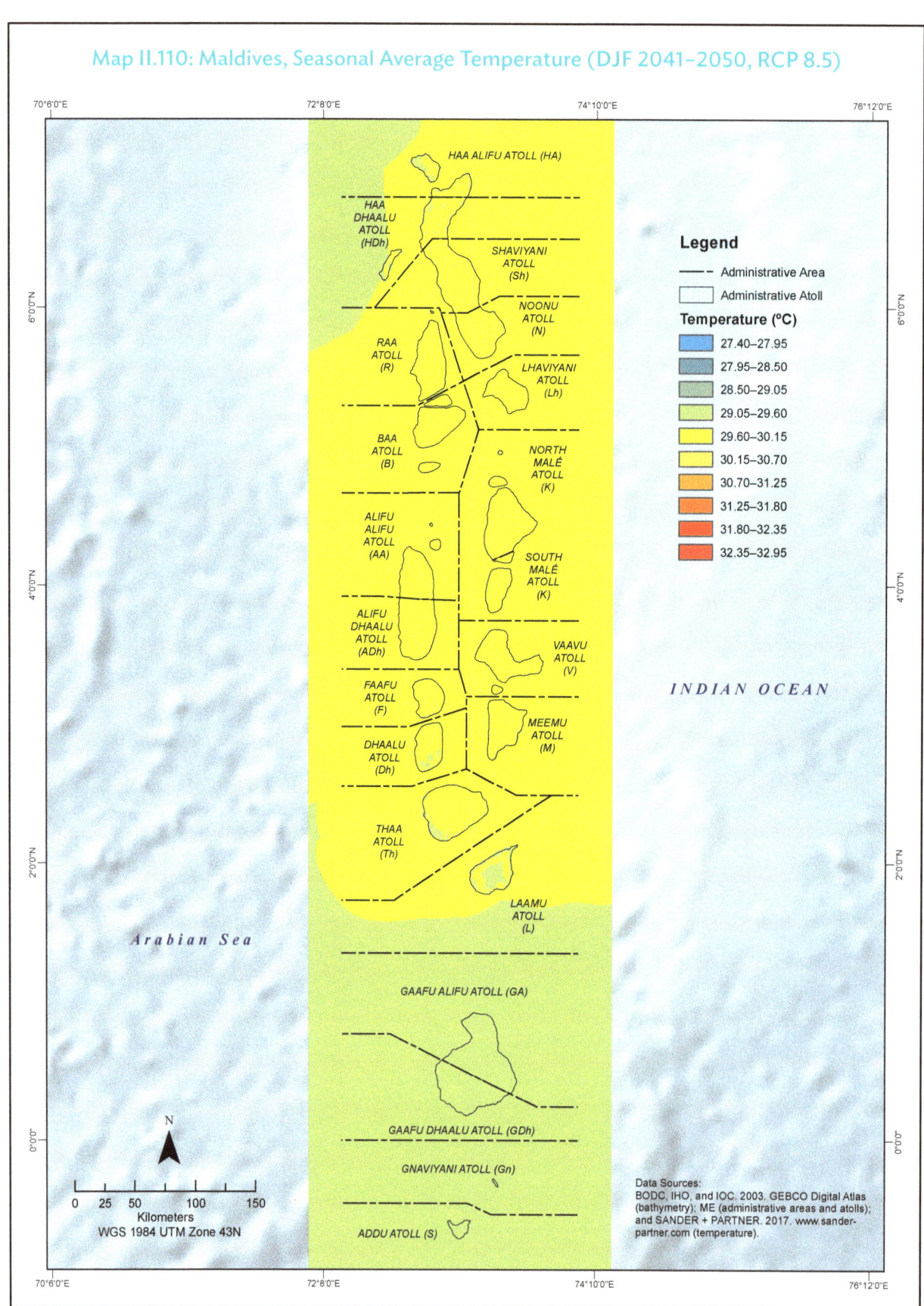

Map II.110: Maldives, Seasonal Average Temperature (DJF 2041–2050, RCP 8.5)

Average Seasonal Temperature Projection (MAM, RCP 8.5)

117

The RCP 8.5 average temperature projection shows that MAM in Maldives will be warmer at 29.60°C–30.15°C in the 2020s and 2030s. Northern Maldives will experience warmer temperature (reaching 30.15°C–30.70°C in the 2040s) during these months.

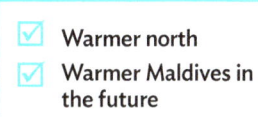

- Warmer north
- Warmer Maldives in the future

Map II.111: Maldives, Average Seasonal Temperature Projection (MAM, RCP 8.5)

Legend
- Administrative Area
- Administrative Atoll

Temperature (°C)
- 27.40–27.95
- 27.95–28.50
- 28.50–29.05
- 29.05–29.60
- 29.60–30.15
- 30.15–30.70
- 30.70–31.25
- 31.25–31.80
- 31.80–32.35
- 32.35–32.95

Data Sources: BODC, IHO, and IOC. 2003. GEBCO Digital Atlas (bathymetry); ME (administrative areas and atoll); and SANDER + PARTNER. 2017. www.sander-partner.com (temperature).

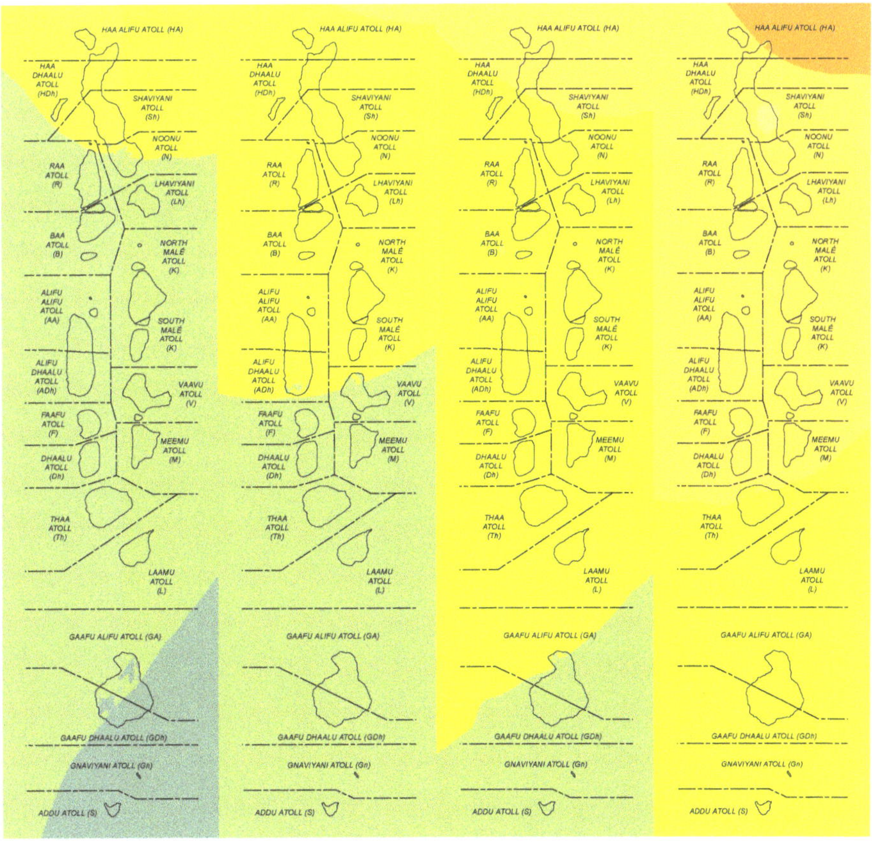

2011–2020 | 2021–2030 | 2031–2040 | 2041–2050

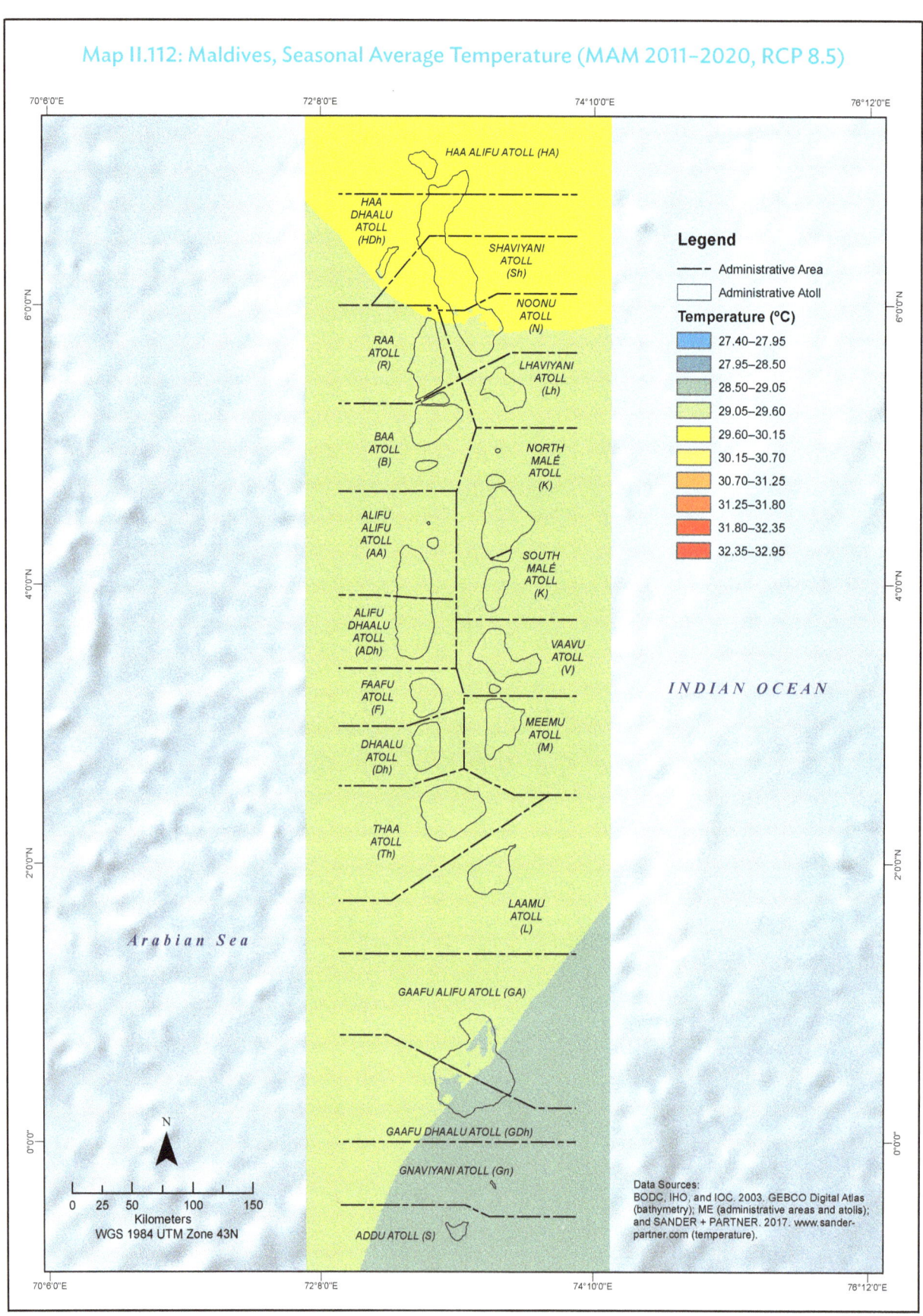

Map II.112: Maldives, Seasonal Average Temperature (MAM 2011–2020, RCP 8.5)

Future Climate 119

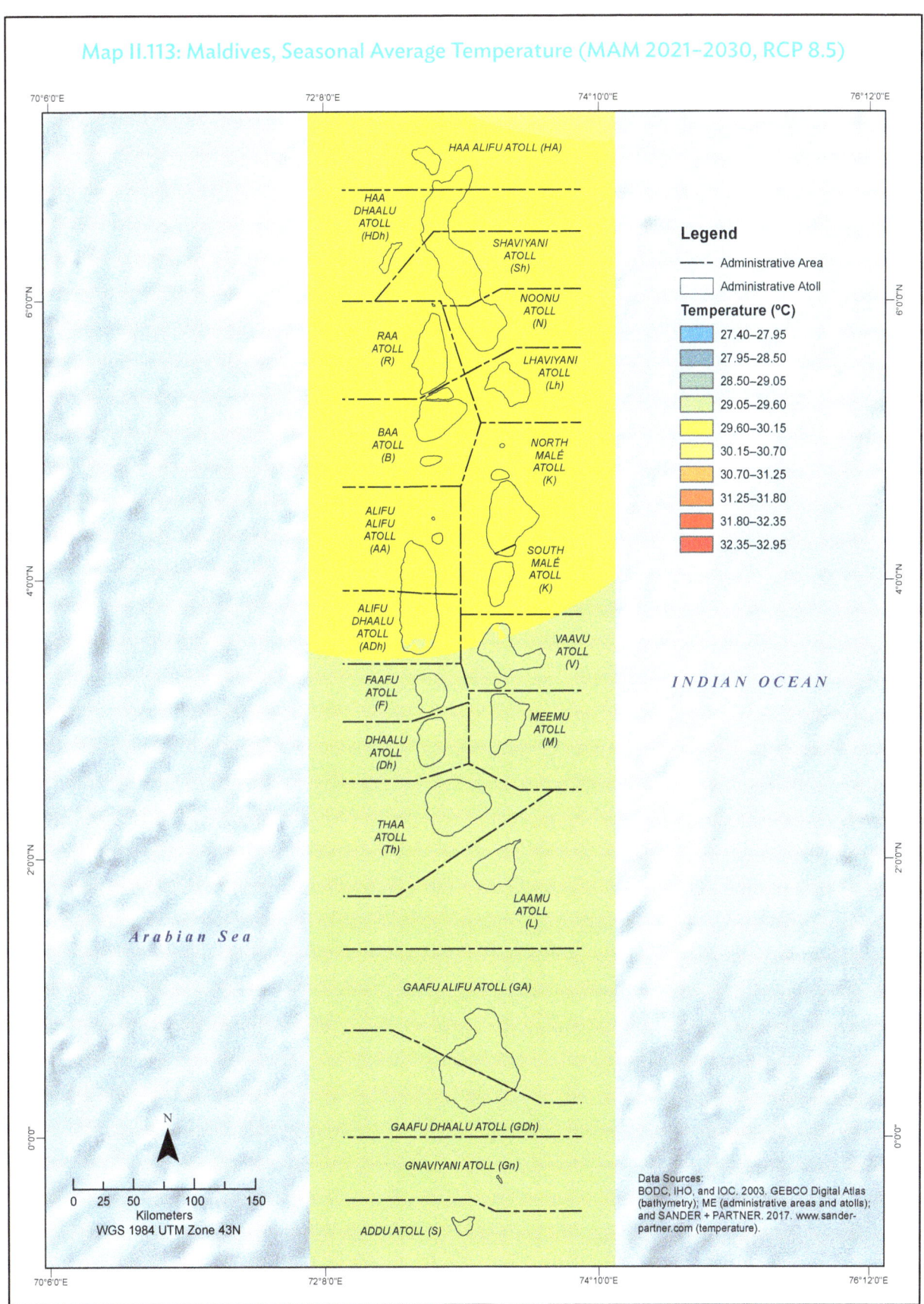

Map II.113: Maldives, Seasonal Average Temperature (MAM 2021–2030, RCP 8.5)

120 Multihazard Risk Atlas of Maldives—Climate and Geophysical Hazards

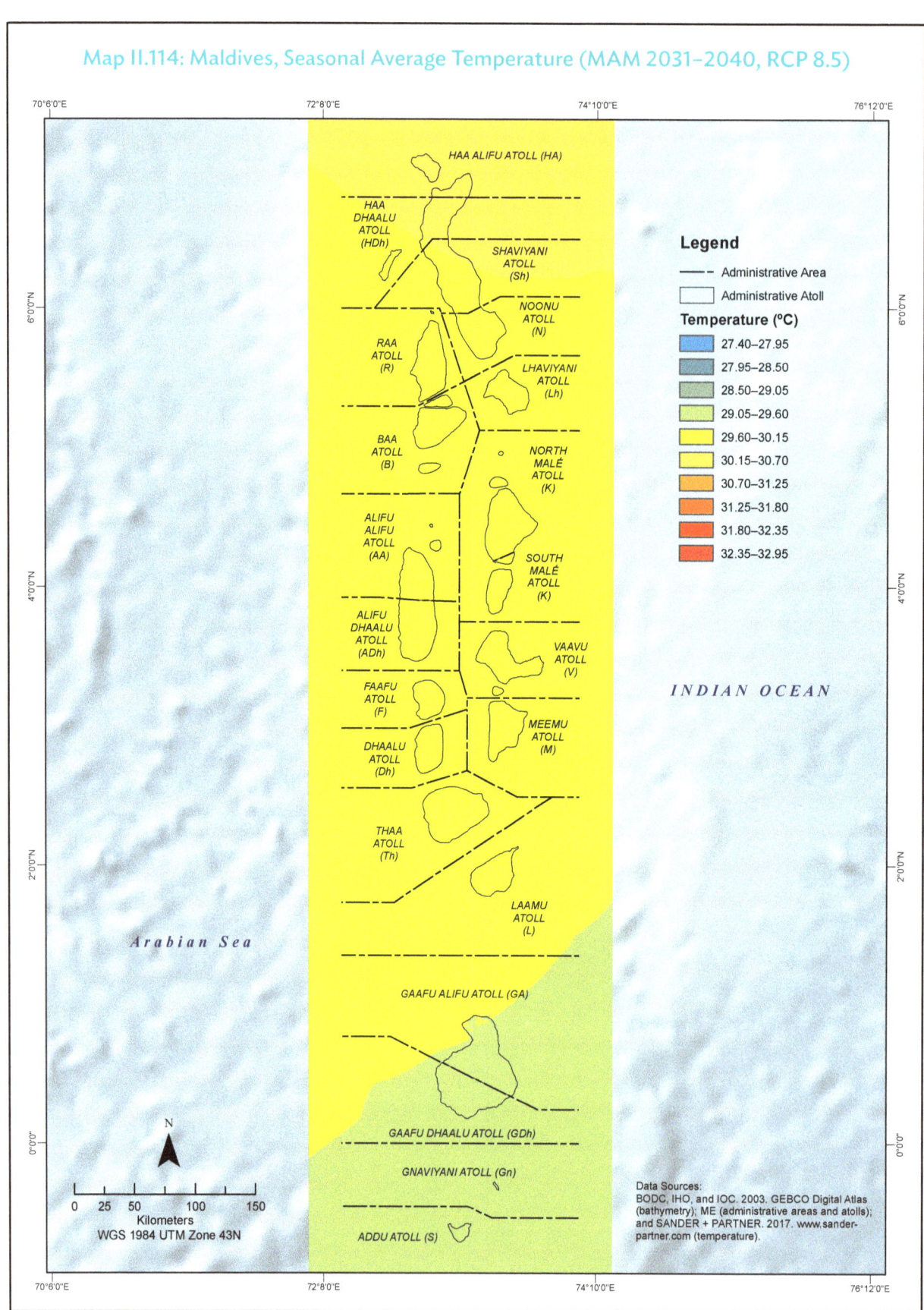

Map II.114: Maldives, Seasonal Average Temperature (MAM 2031–2040, RCP 8.5)

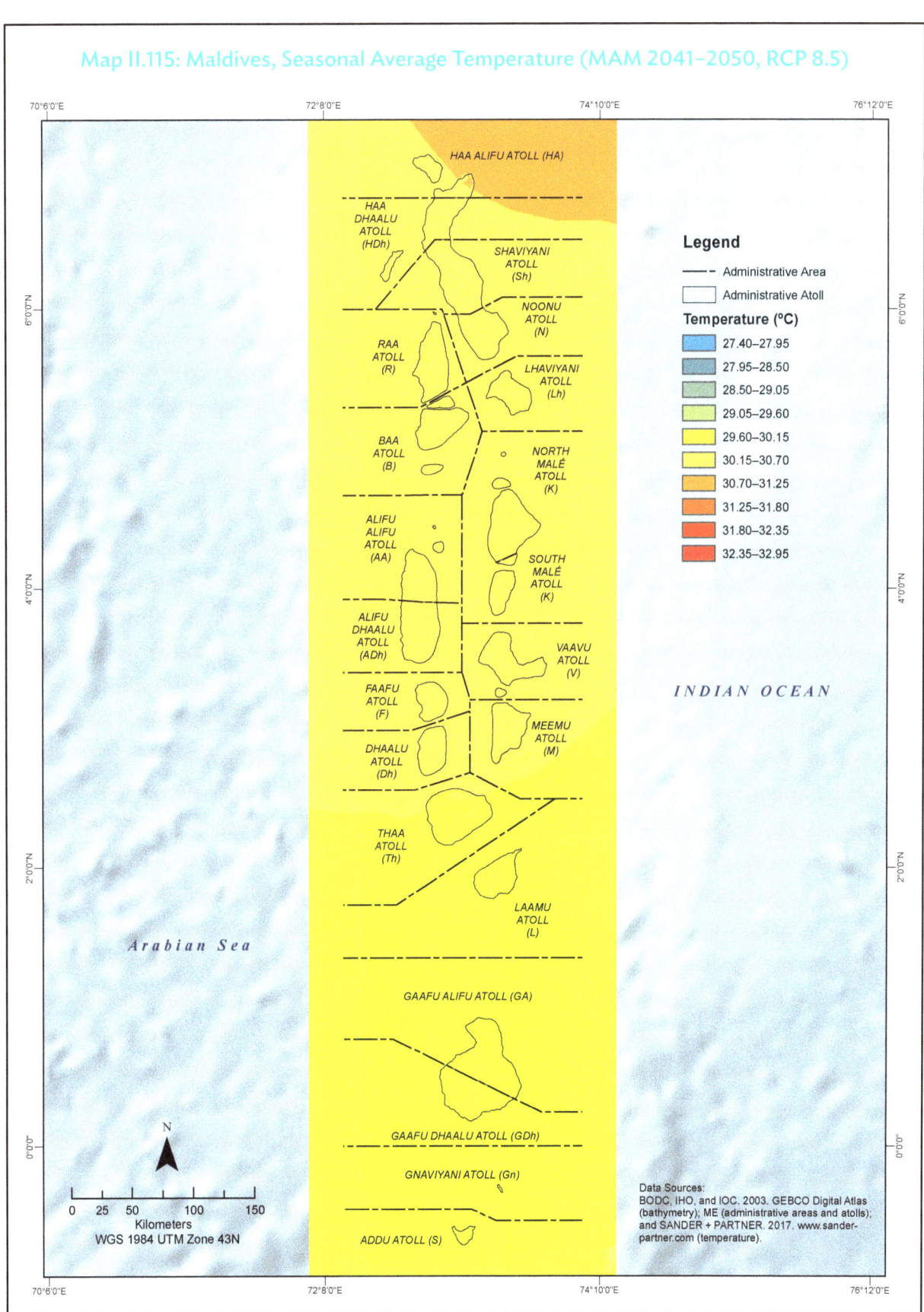
Map II.115: Maldives, Seasonal Average Temperature (MAM 2041–2050, RCP 8.5)

Average Seasonal Temperature Projection (JJA, RCP 8.5)

- ☑ Warmer north
- ☑ Warming in middle Maldives in the future
- ☑ RCP 8.5 projected higher temperature in 2041–2050 than RCP 4.5

For JJA, the pattern of average seasonal projected temperature based on RCP 8.5 shows that the northern and middle portions of Maldives are warmer compared to the southern region. JJA will be warmer in the coming decades, especially in the 2040s, as atolls experience temperatures ranging from 29.05°C to 30.15°C, with higher temperatures in the middle portion of Maldives. As with other seasonal projections, RCP 8.5 depicts a warmer scenario compared to RCP 4.5.

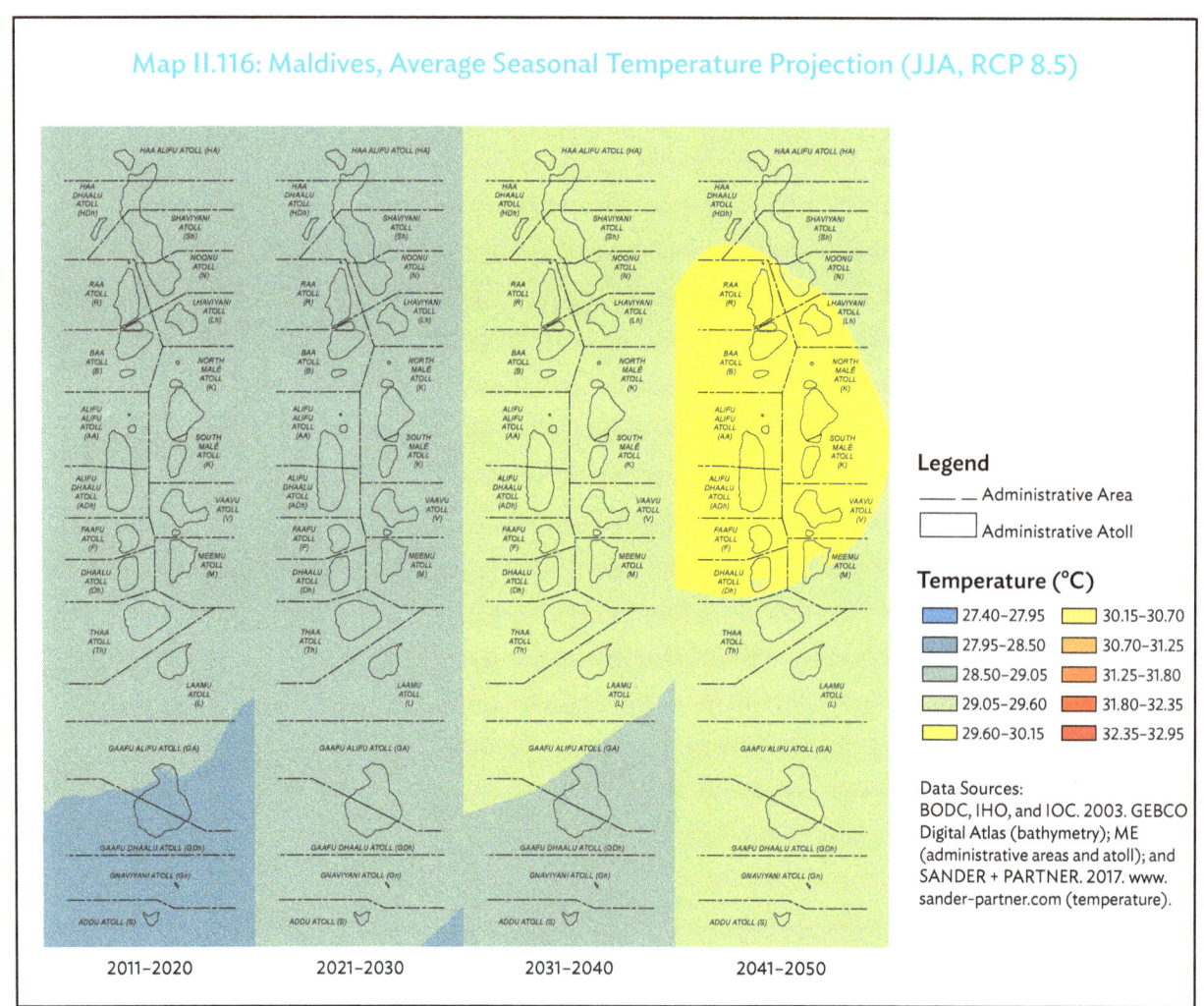

Map II.116: Maldives, Average Seasonal Temperature Projection (JJA, RCP 8.5)

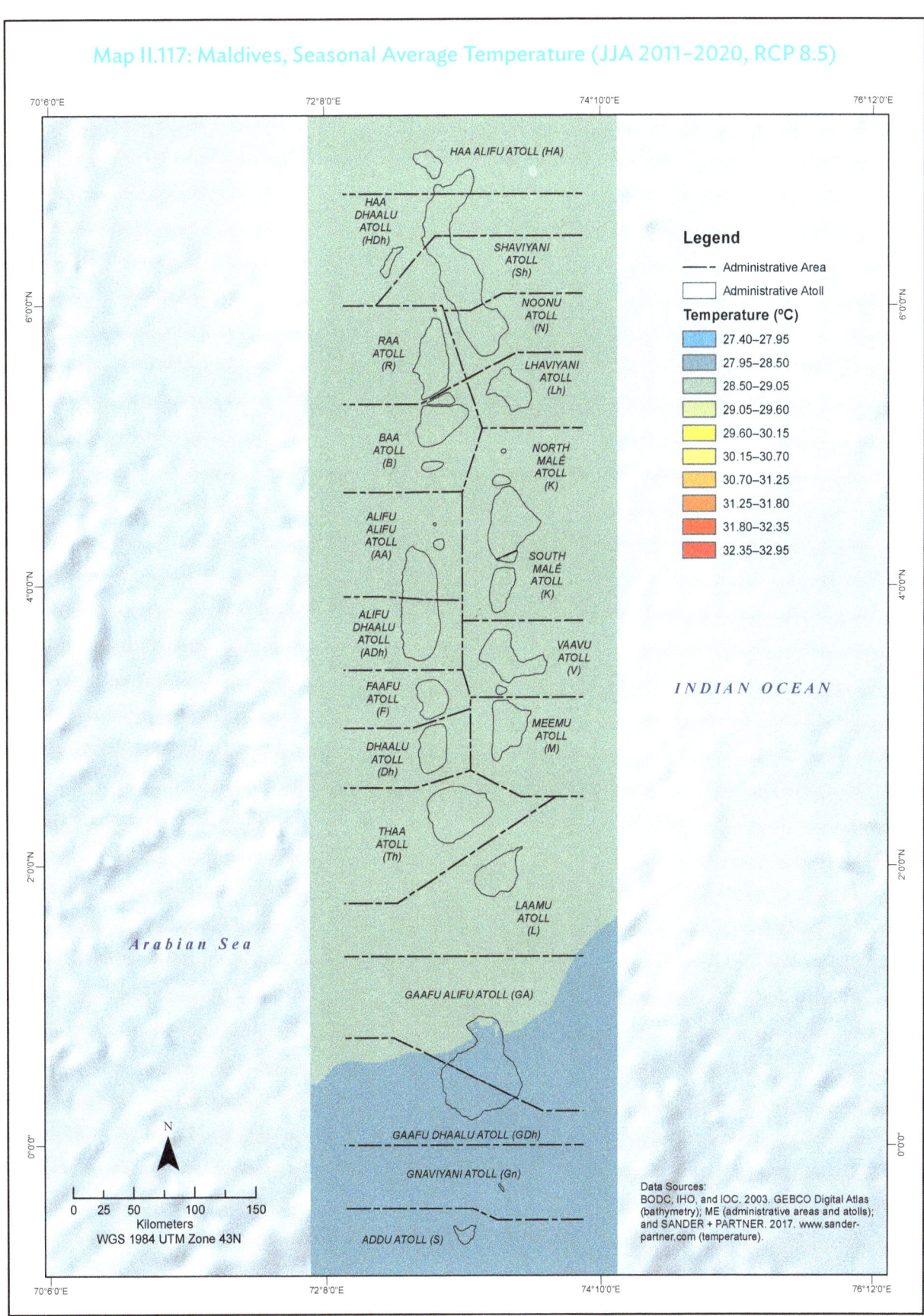

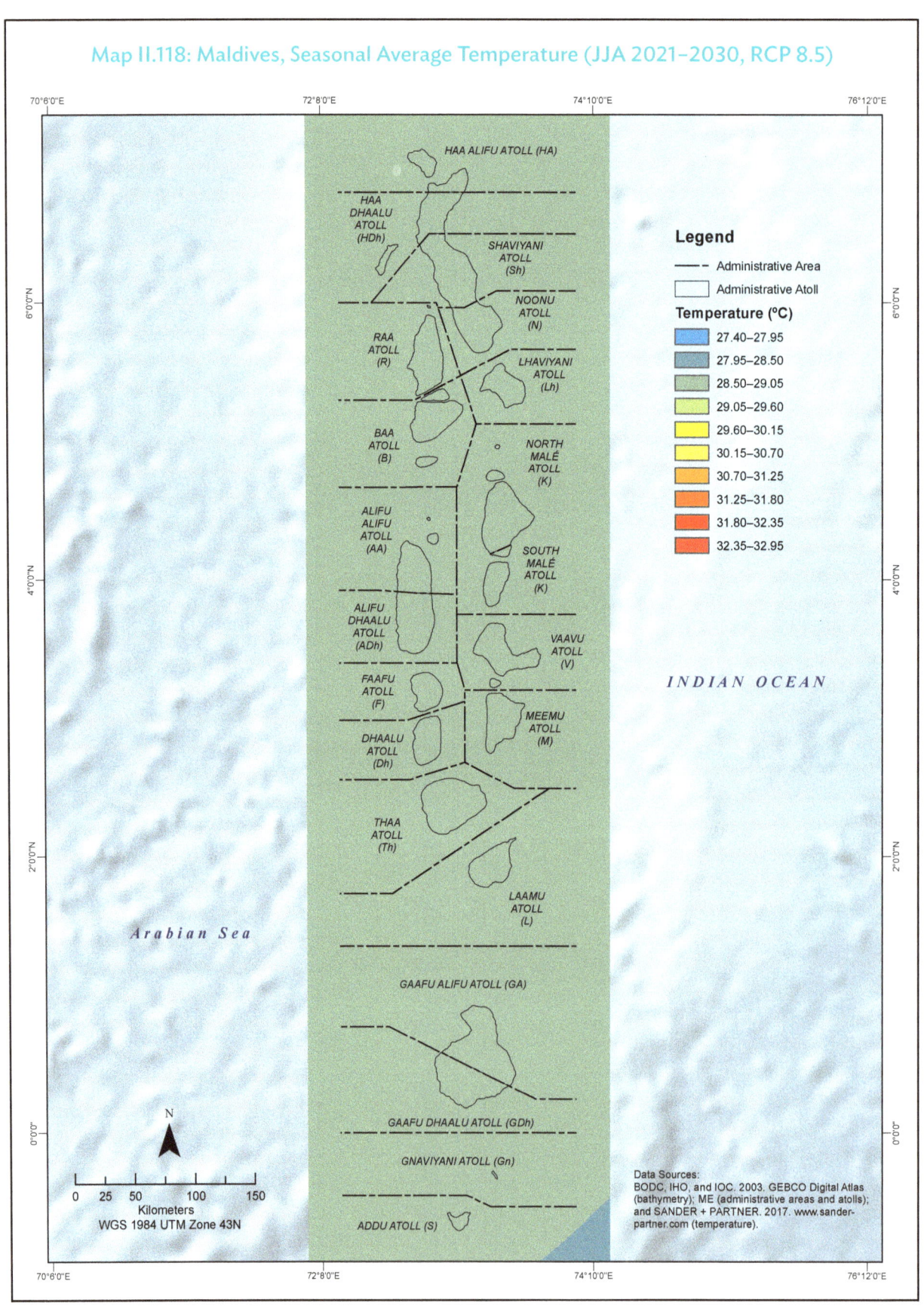

Map II.118: Maldives, Seasonal Average Temperature (JJA 2021–2030, RCP 8.5)

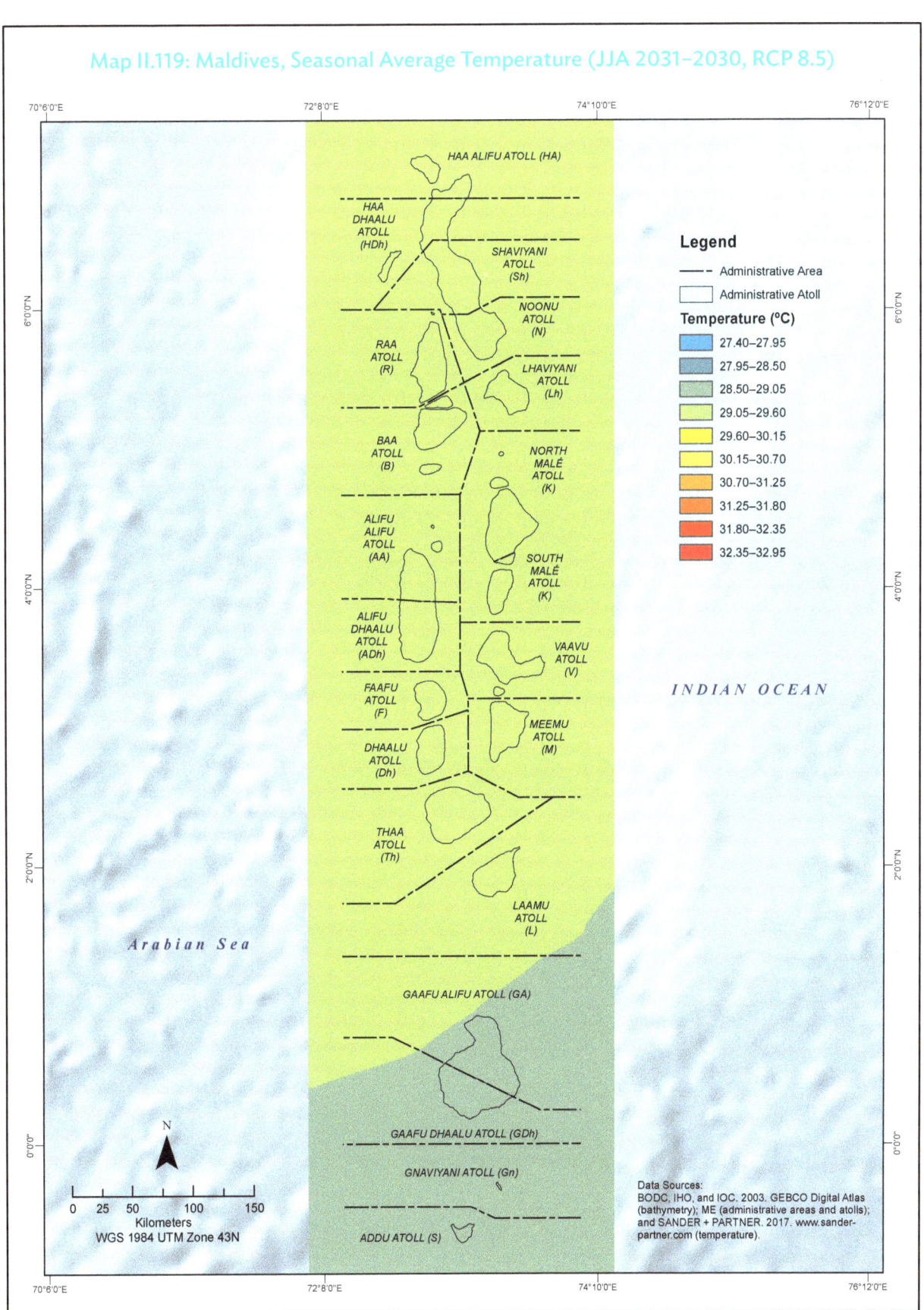

Map II.119: Maldives, Seasonal Average Temperature (JJA 2031–2030, RCP 8.5)

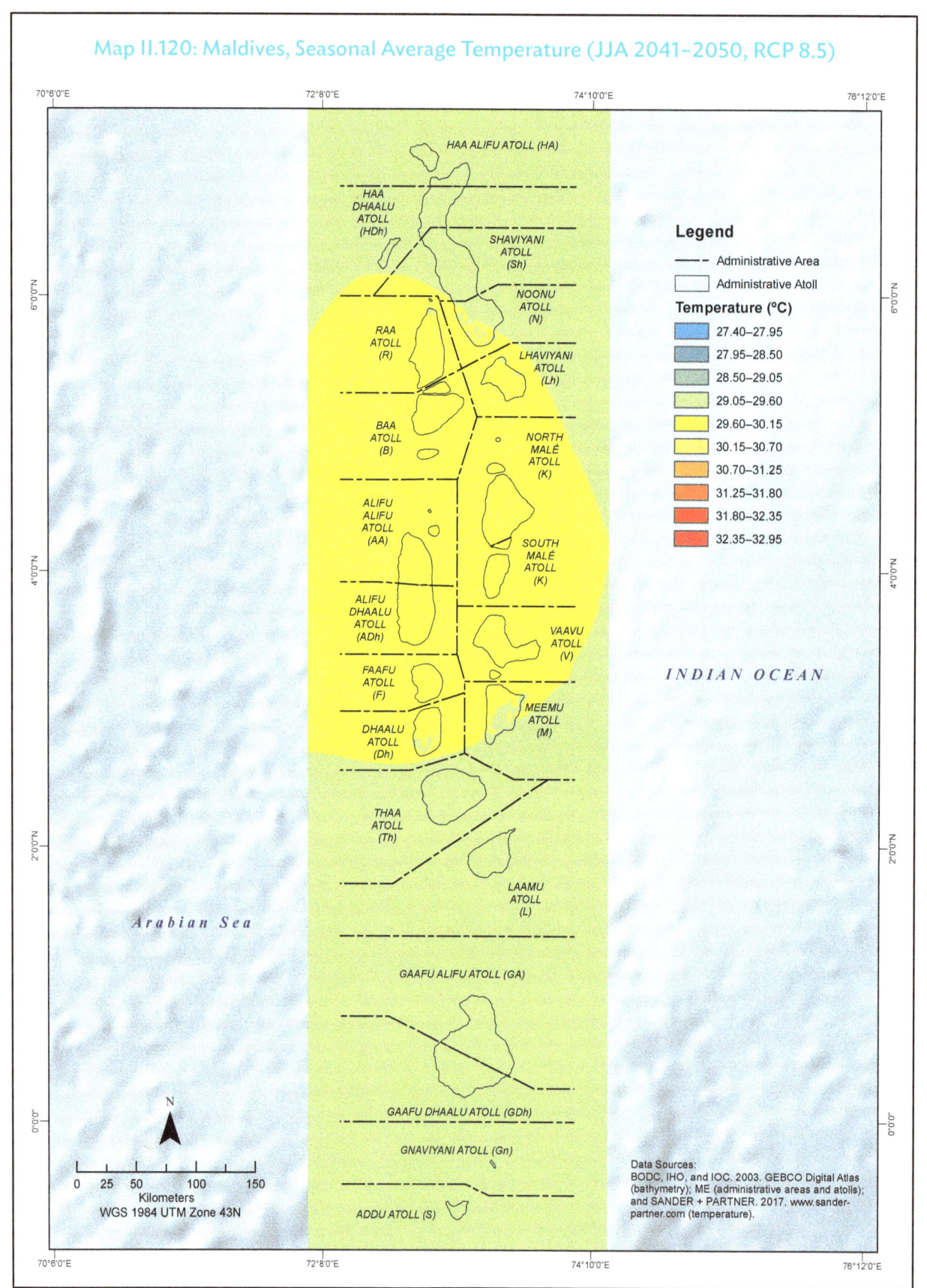

Map II.120: Maldives, Seasonal Average Temperature (JJA 2041–2050, RCP 8.5)

Average Seasonal Temperature Projection (SON, RCP 8.5)

SON will still be the coldest months, according to the average temperature projection using RCP 8.5. The northern half of Maldives will be warmer in the 2020s and 2040s, compared to the 2010s and 2030s when the temperature is even from north to south. The coming decades will be warmer as the temperature increases from 27.95°C–28.50°C in the 2010s to 29.05°C–29.60°C in the 2040s.

- ☑ Warmer north (2020s and 2040s)
- ☑ Warming in the future
- ☑ RCP 8.5 projected higher temperature in 2041–2050 than RCP 4.5

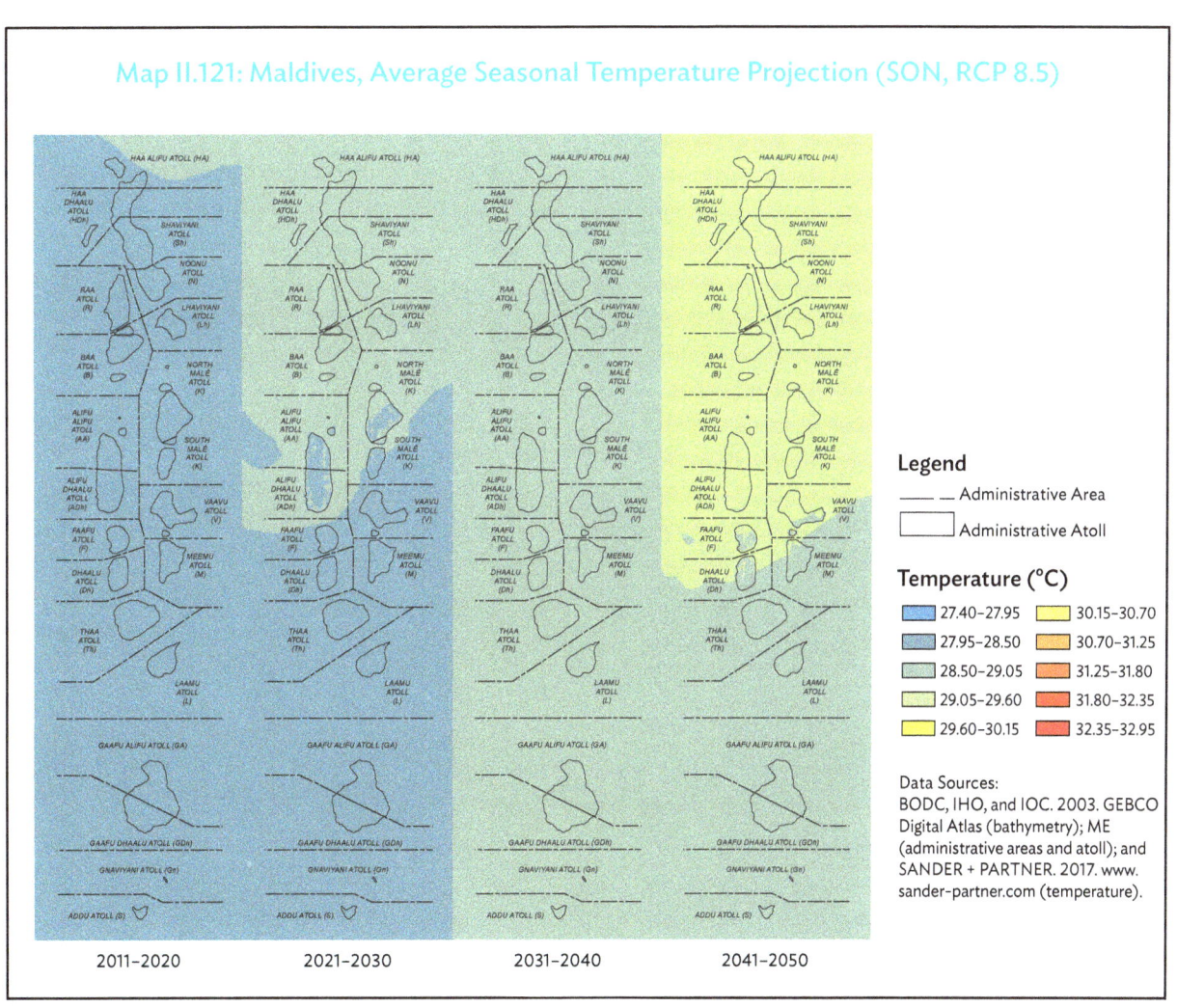

Map II.121: Maldives, Average Seasonal Temperature Projection (SON, RCP 8.5)

Data Sources: BODC, IHO, and IOC. 2003. GEBCO Digital Atlas (bathymetry); ME (administrative areas and atoll); and SANDER + PARTNER. 2017. www.sander-partner.com (temperature).

128 Multihazard Risk Atlas of Maldives—Climate and Geophysical Hazards

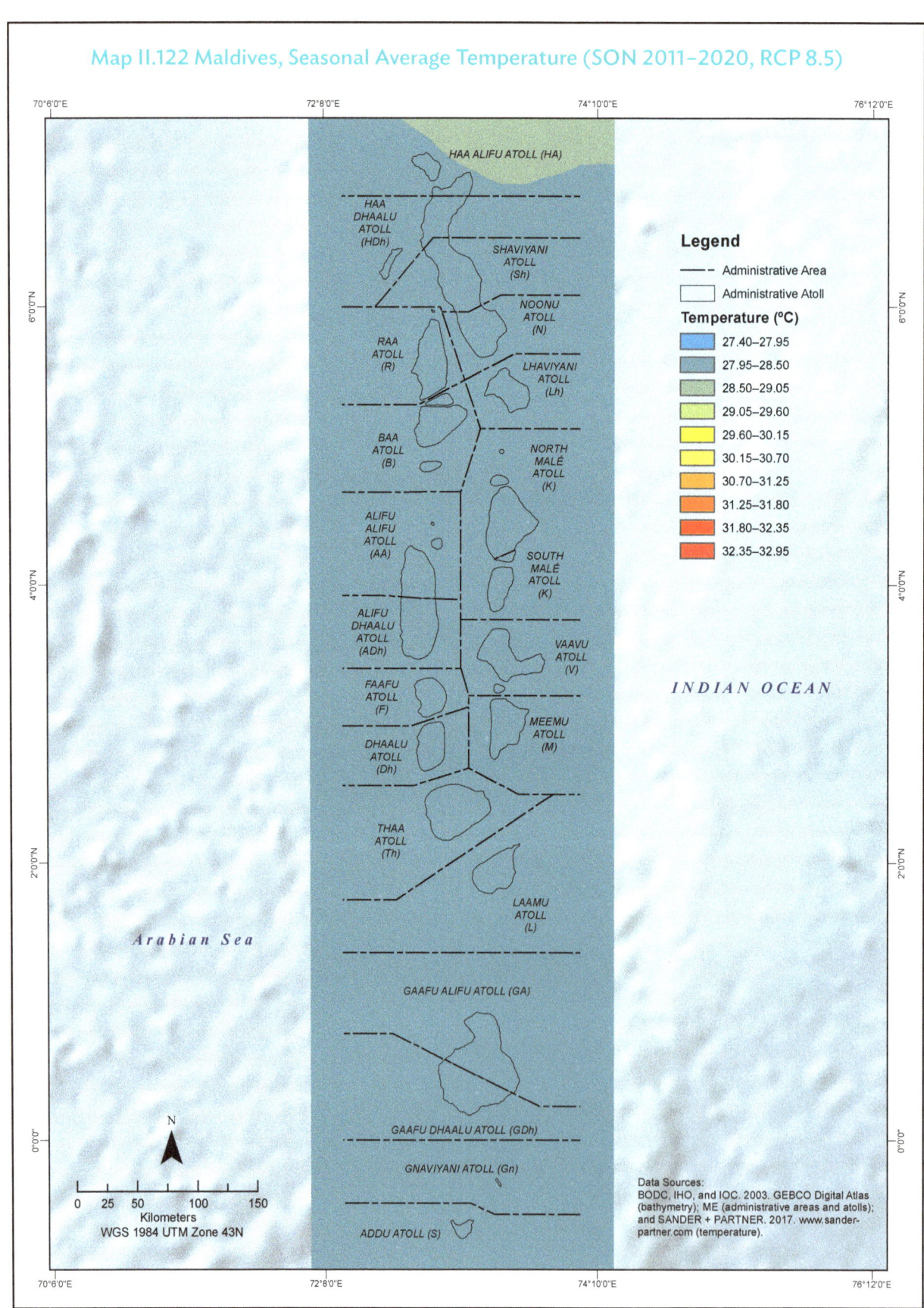

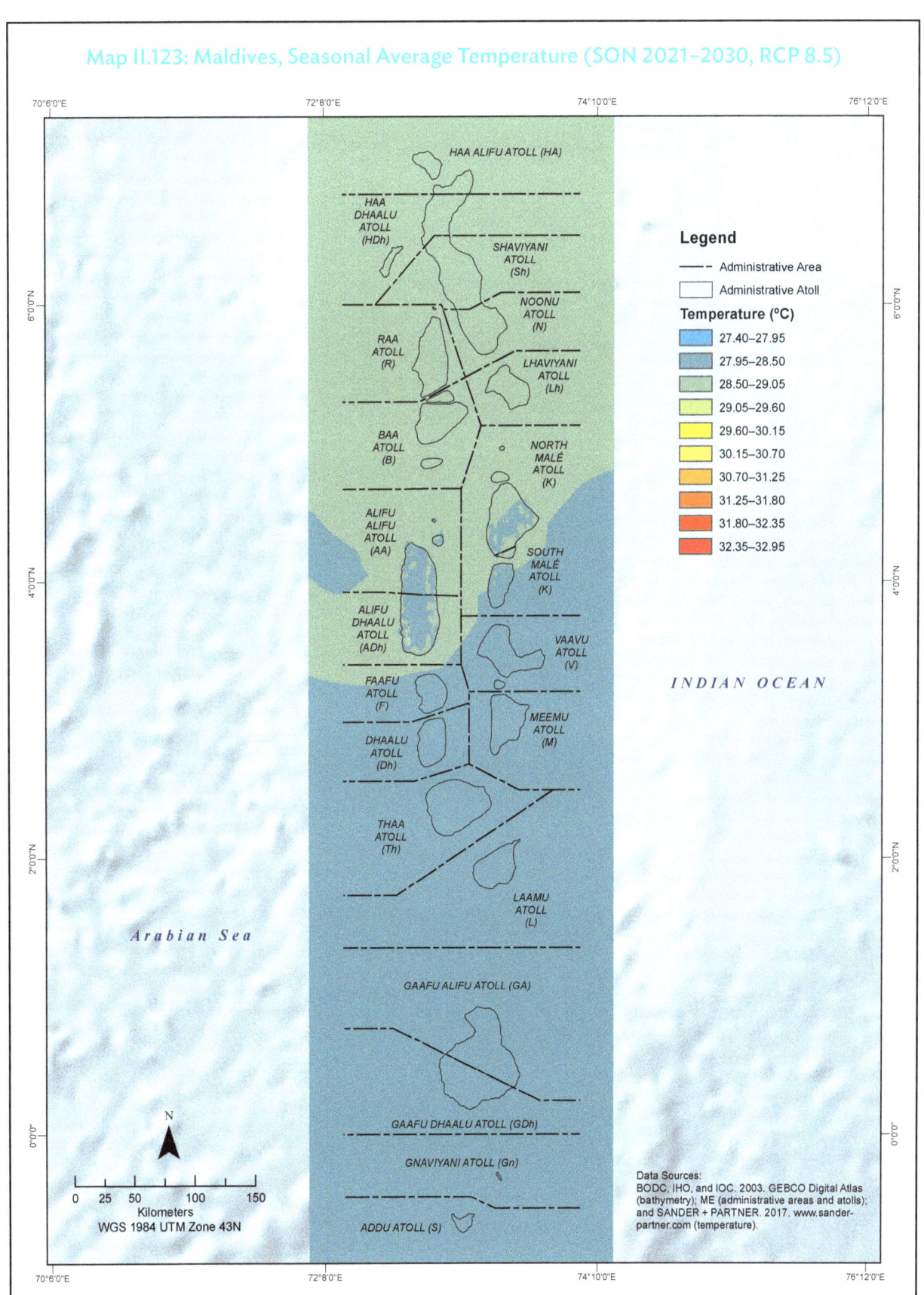

Map II.123: Maldives, Seasonal Average Temperature (SON 2021–2030, RCP 8.5)

130 Multihazard Risk Atlas of Maldives—Climate and Geophysical Hazards

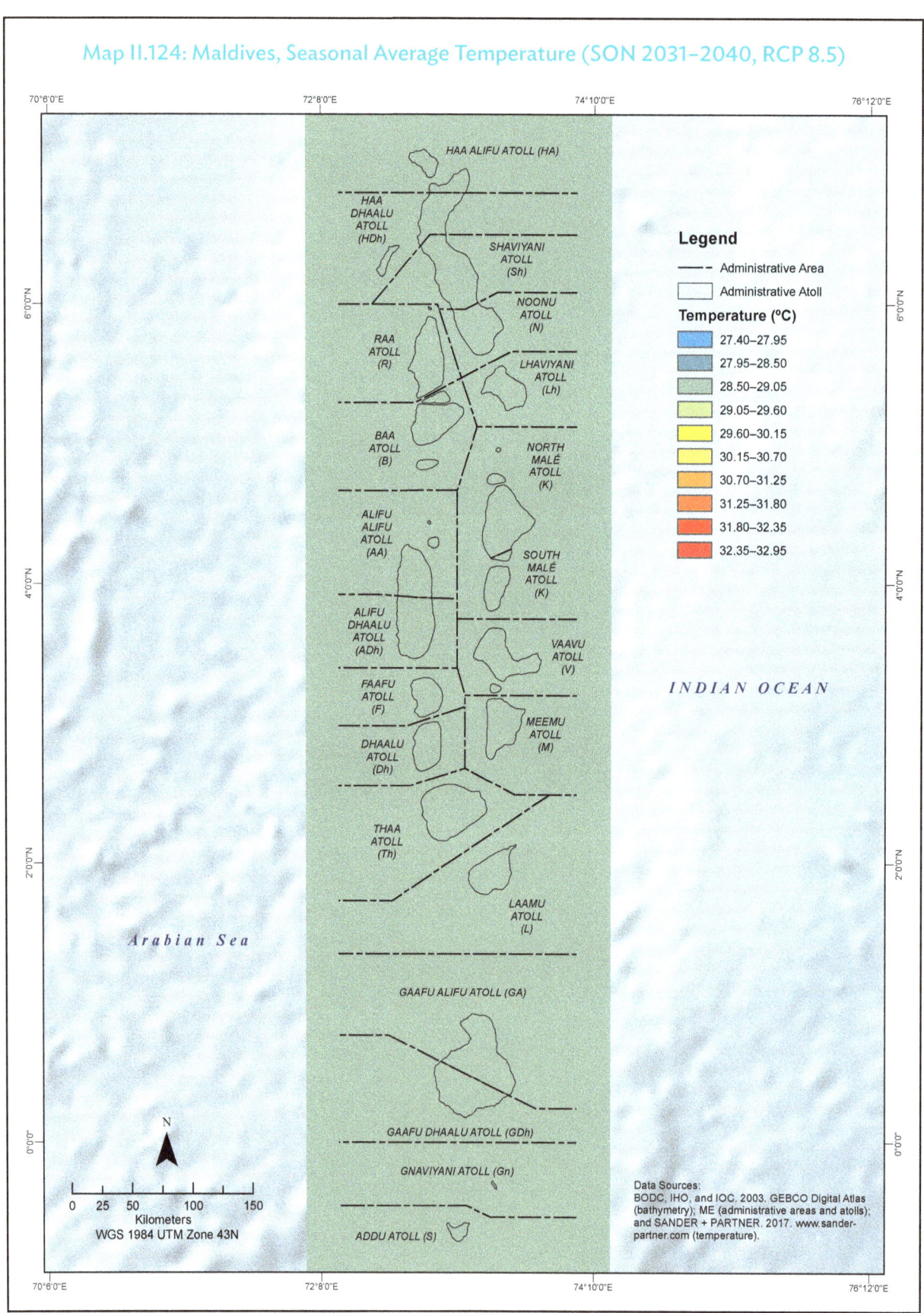

Map II.124: Maldives, Seasonal Average Temperature (SON 2031–2040, RCP 8.5)

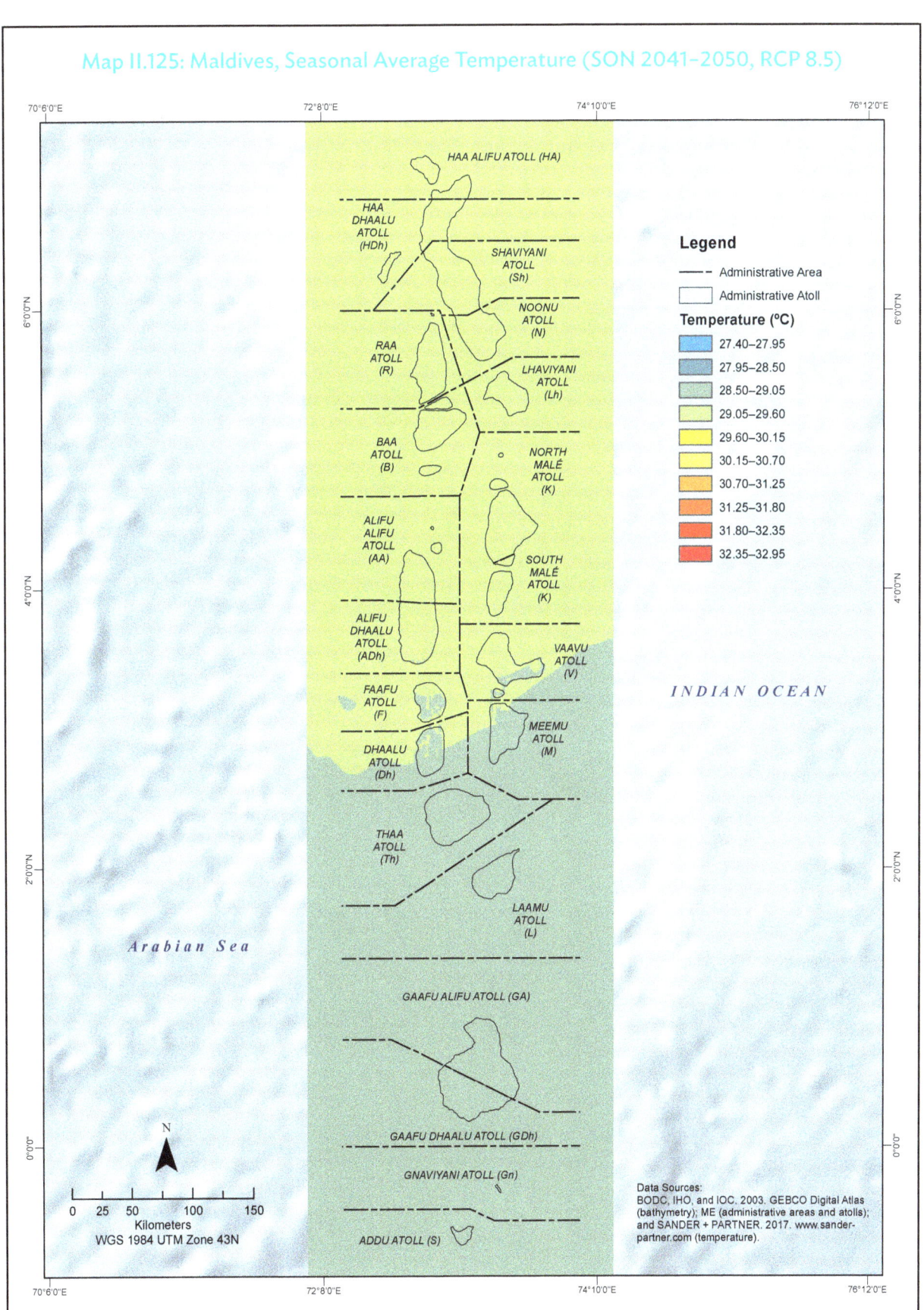

Map II.125: Maldives, Seasonal Average Temperature (SON 2041–2050, RCP 8.5)

Summary of Observations for Rainfall

Similar to the historical annual average rainfall distribution, northern Maldives will continue to receive more rainfall compared to the southern region. However, this may vary depending on the season. The wettest months will still be JJA. During this period, rainfall will mostly be distributed in the north. Except for the slight variation in the area receiving 4–8 mm/day of rainfall, the projections for RCP 4.5 and RCP 8.5 show no difference.

Table II.2: Variation in Rainfall Patterns across Space, Time, and Representative Concentration Pathways

Climate Maps	Spatial	Seasonal	Decadal	RCP 4.5 vs RCP 8.5
Average Annual	Wetter north (4.6–5.8 mm/day)	-	-	-
Average Seasonal	Wetter north	Generally dry season	-	-
Projected Mean Annual (4.5)	Wetter north	-	Wetter north in the future	No difference with 8.5
Projected Mean Seasonal (4.5) DJF	Wetter south	Dry season	Wetter southern atolls	Smaller coverage of 4–8 mm/day
Projected Mean Seasonal (4.5) MAM	Generally dry	Driest season	No change	No difference with 8.5
Projected Mean Seasonal (4.5) JJA	Wetter north	Wettest season	Wetter north in the future	No difference with 8.5
Projected Mean Seasonal (4.5) SON	Wetter north	Wet season	No change	Smaller coverage of 4–8 mm/day
Projected Mean Annual (8.5)	Wetter north	-	Wetter north in the future	No difference with 4.5
Projected Mean Seasonal (8.5) DJF	Wetter south	Dry season	More atolls in the south will experience higher rainfall rate	Greater coverage of 4–8 mm/day
Projected Mean Seasonal (8.5) MAM	Generally dry	Driest season	No change	No difference with 4.5
Projected Mean Seasonal (8.5) JJA	Wetter north	Wettest season	Wetter north in the future	No difference with 4.5
Projected Mean Seasonal (8.5) SON	Wetter north	Wet season	More atolls in the north with 4–8 mm/day rainfall rate	Greater coverage of 4–8 mm/day

– = not applicable; DJF = December, January, February; JJA = June, July, August; MAM = March, April, May; mm/day = millimeter per day; RCP = representative concentration pathway; SON = September, October, November.

Source: SANDER + PARTNER. 2017. www.sander-partner.com.

Summary of Observations for Temperature

Maldives will experience a warmer climate in the 2030s and 2040s. Greater warming will be felt in the northern atolls, especially in MAM. This pattern is consistent across seasons and decades. However, the RCP 8.5 projections show a significantly warmer possibility compared with the RCP 4.5.

Table II.3: Variation in Temperature Patterns across Space, Time, and Representative Concentration Pathways

Climate Maps	Spatial	Seasonal	Decadal	RCP 4.5 vs RCP 8.5
Average Annual	Warmer north (28–28.4°C)	–	Warmer Maldives especially in the future	–
Average Seasonal	–	MAM is the warmest month, SON is the coldest month	–	–
Projected Mean Annual (4.5)	Warmer north	–	Future warmer Maldives especially in the north	Lower temperature compared to 8.5
Projected Mean Seasonal (4.5) DJF	Warmer north	–	Warmer Maldives especially in the future	Lower temperature compared to 8.5
Projected Mean Seasonal (4.5) MAM	Warmer north	MAM is the warmest month	Future warmer Maldives especially in the north	Lower temperature compared to 8.5
Projected Mean Seasonal (4.5) JJA	Warmer north	–	Future warmer Maldives especially in the north	Lower temperature compared to 8.5
Projected Mean Seasonal (4.5) SON	Warmer north	SON is the coldest month	Future warmer Maldives especially in the north	Lower temperature compared to 8.5
Projected Mean Annual (8.5)	Warmer north	–	Future warmer Maldives especially in the north	Higher temperature compared to 4.5
Projected Mean Seasonal (8.5) DJF	Even temperature across Maldives except in 2040s	–	Future warmer Maldives especially in the north	Higher temperature compared to 4.5
Projected Mean Seasonal (8.5) MAM	Warmer north	Warmest months	Future warmer Maldives especially in the north	Higher temperature compared to 4.5
Projected Mean Seasonal (8.5) JJA	Warmer north	–	Warming in the middle of Maldives	Higher temperature compared to 4.5
Projected Mean Seasonal (8.5) SON	Warmer north in 2020s and 2040s	Coldest months	Future warmer Maldives especially in the north	Higher temperature compared to 4.5

– = not applicable; °C = degree Celsius; DJF = December, January, February; JJA = June, July, August; MAM = March, April, May; RCP = representative concentration pathway; SON = September, October, November.

Source: SANDER + PARTNER. 2017. www.sander–partner.com.

Geophysical Hazards

Maldives has experienced several disasters in the past. According to a 2006 United Nations Development Programme (UNDP) report in which cyclone tracks were analyzed across the century, Maldives has only experienced 11 cyclones in 12 decades, with the strongest cyclones crossing northern Maldives (UNDP 2006). Although cyclones are not a frequently recurring hazard in the country, they still bring wind, rain, and storm surges to the low-lying islands. Flooding, usually caused by a surge, has recurred several times in Maldives. *Bodu raalhu* (big waves) flooded a few islands in 1987. Northeastern islands are mostly affected by storm surges (UNDP 2006).

Aside from cyclones, Maldives also experiences earthquakes and tsunamis. In 25 years, three earthquakes with a magnitude of at least 7.0 hit Maldives (UNDP 2006). The UNDP report also estimated the decay of peak ground acceleration for a 475-year return period. Results showed that the southern Gnaviyani Atoll and Addu Atoll have the highest peak ground acceleration. Movements at subduction zones, usually occurring at the edges of the tectonic plates, generate earthquakes, which in turn generate tsunamis. The most devastating disaster was the tsunami on 26 December 2004. It generated waves up to 4.2 meters that damaged at least a dozen of the inhabited islands and thousands of homes, affected thousands of lives, and lost about two-thirds of the country's gross domestic product (UNDP 2006). Based on UNDP's model, the eastern borders of atolls have higher probability of experiencing tsunamis at 320–450 centimeters high.

Tsunami Monument. Various hazards, including tsunamis, have affected Maldives in the past. This Tsunami Monument stands at the shores of Malé in memory of the people who died during the 2004 tsunami (photo by Utsav Mulay).

Geophysical Hazards   135

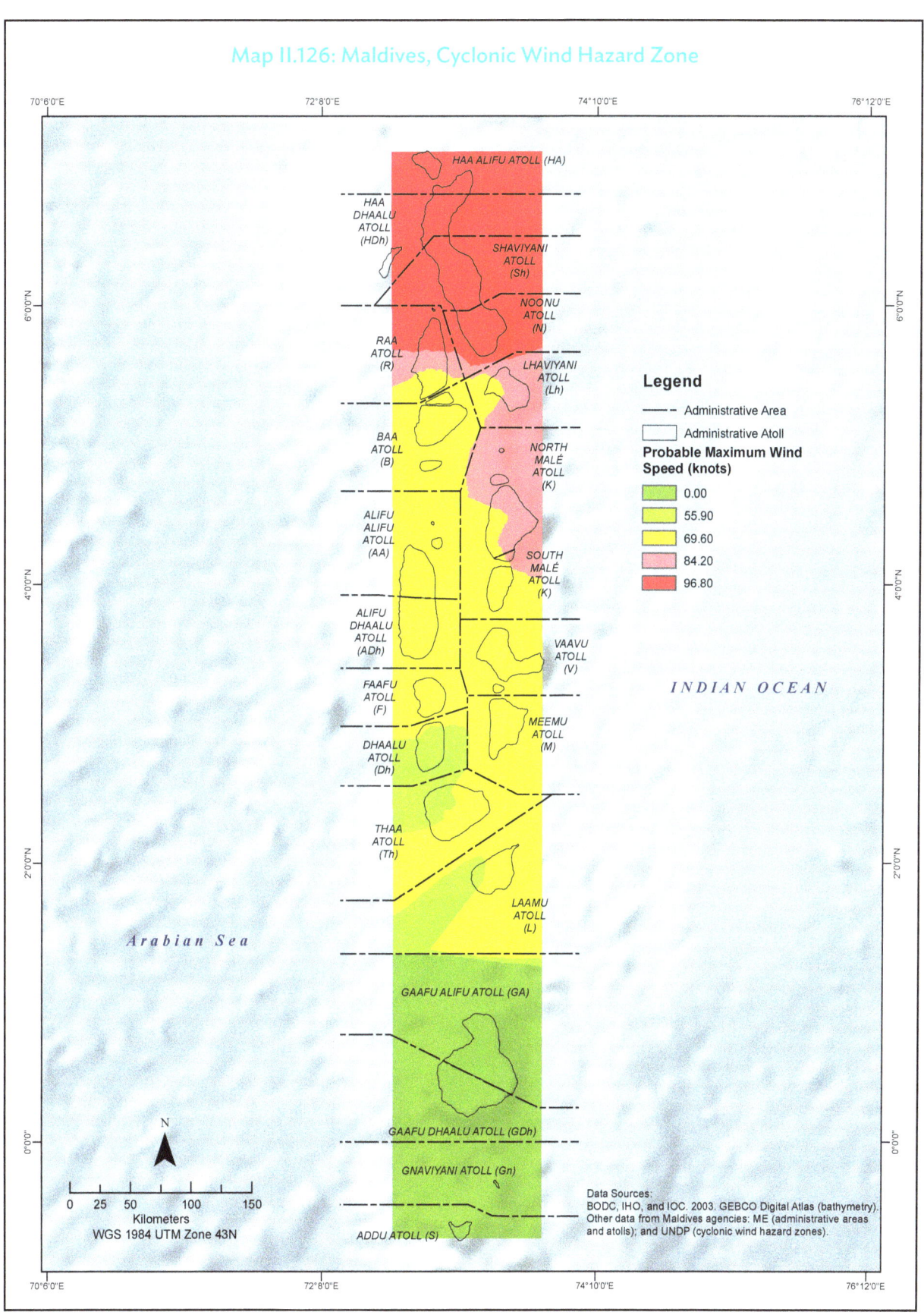

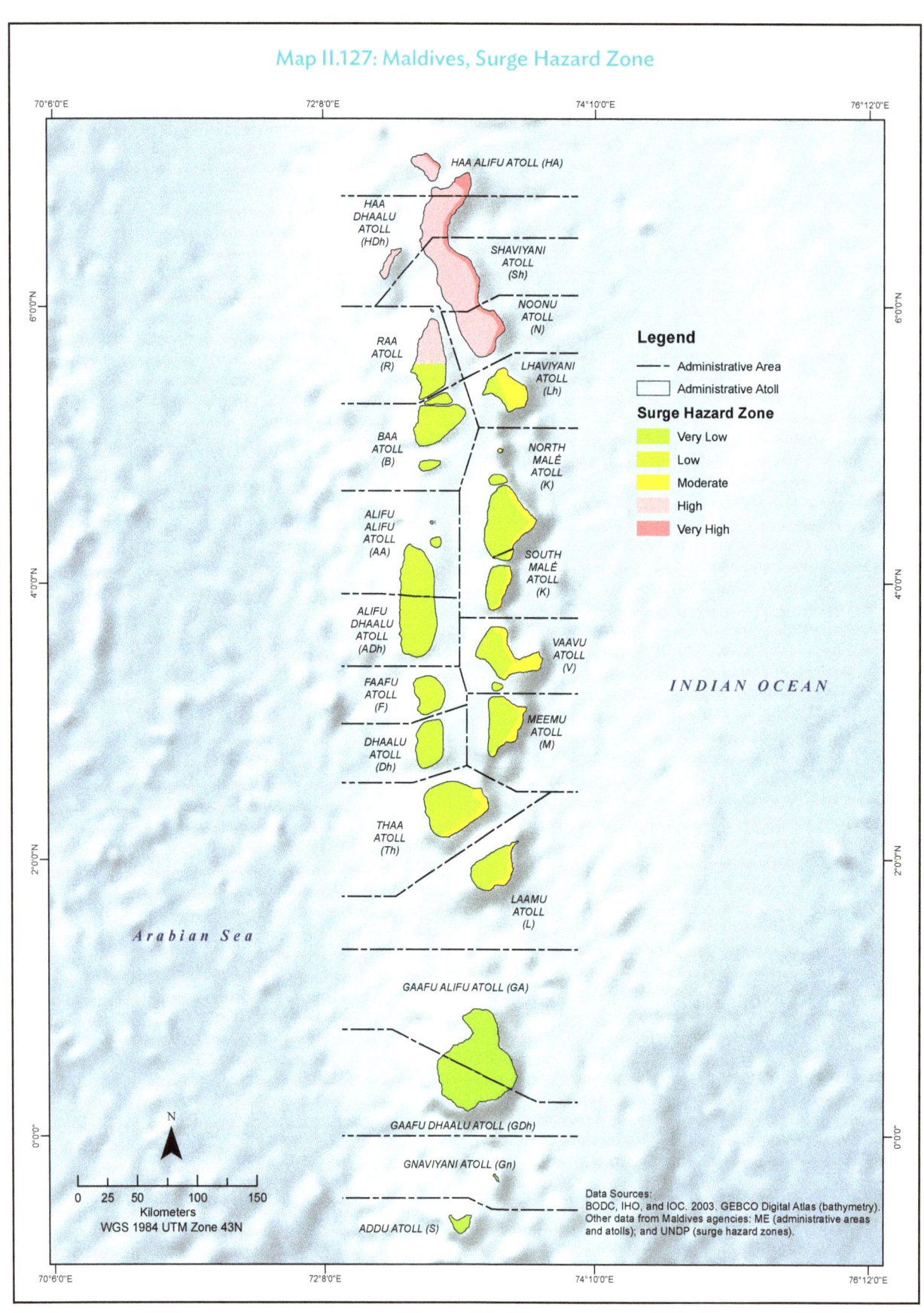

Geophysical Hazards 137

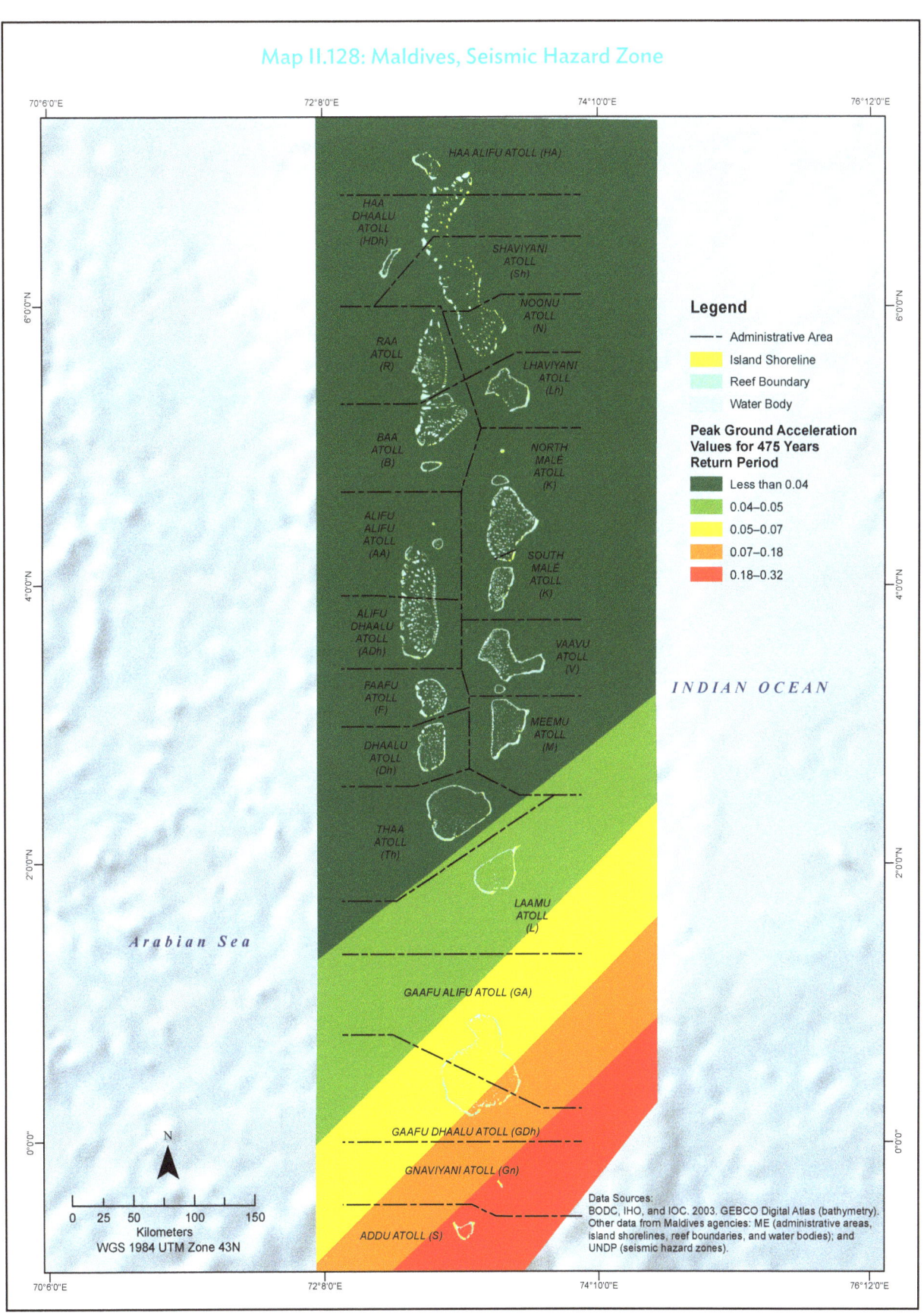

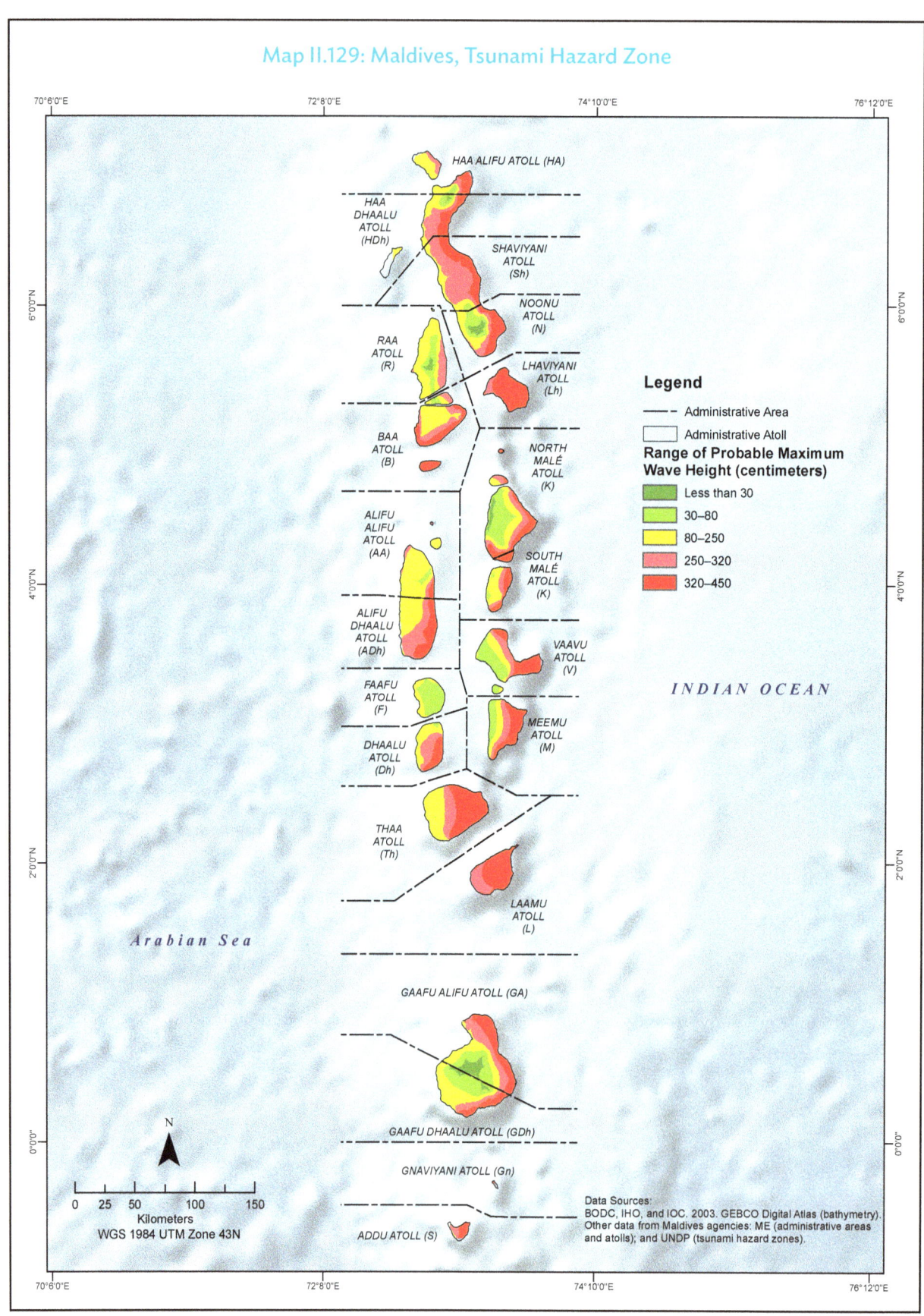

Map Data Sources

Government Ministries, Departments, and Agencies in Maldives
    Civil Aviation Authority
        Airports
    Land and Survey Authority
        Atoll capital islands
        Cities
    Meteorological Service
        Automatic weather stations
        Meteorological observation stations
    Ministry of Economic Development
        Ports
    Ministry of Environment
        Administrative areas
        Administrative atolls
        Island shorelines
        Reef boundaries
        Water bodies

International Institutions
    SANDER + PARTNER. www.Sander-Partner.com
        Rainfall
        Temperature

References

Asian Development Bank. 2015. *Maldives: Overcoming the Challenges of a Small Island State. Country Diagnostic Study.* Manila.

British Oceanographic Data Centre (BODC), International Hydrographic Organisation (IHO) and the Intergovernmental Oceanographic Commission (IOC) of the United Nations Educational, Scientific and Cultural Organization. 2003. *General Bathymetric Chart of the Oceans (GEBCO) Digital Atlas.* UK: British Oceanographic Data Centre.

Intergovernmental Panel on Climate Change (IPCC). 2014. *Climate Change 2014: Synthesis Report. Contribution of Working Groups I, II and III to the Fifth Assessment Report of the Intergovernmental Panel on Climate Change* [Core Writing Team, R.K. Pachauri, and L.A. Meyer, eds.]. Geneva: IPCC.

Khan, T., D. Quadir, T. Murty, A. Kabir, F. Aktar, and M. Sarker. 2002. Relative Sea Level Changes in Maldives and Vulnerability of Land Due to Abnormal Coastal Inundation. *Marine Geodesy* 25 (1–2). pp. 133–143.

Moosa, F. S. 2014. *Country Report: Republic of Maldives.* Asian Disaster Reduction Centre.

United Nations Development Programme. 2006. *Developing a Disaster Risk Profile for Maldives.*

Maps were prepared by the Country Consultant Team and the Manila Observatory on behalf of the Asian Development Bank.

www.ingramcontent.com/pod-product-compliance
Lightning Source LLC
Chambersburg PA
CBHW040546220526
45473CB00017B/3036